AA001132

Stress Induced Phenomena and Reliability in 3D Microelectronics

Printed from e-media with permission by:

Curran Associates, Inc.
57 Morehouse Lane
Red Hook, NY 12571

Some format issues inherent in the e-media version may also appear in this print version.

Printed by Curran Associates, Inc. (2014)
Print on Demand ISBN: (978-1-63266-741-0)

To learn more about the AIP Conference Proceedings Series,
Please visit **http://proceedings.aip.org**

Additional copies of this publication are available from:

Curran Associates, Inc.
57 Morehouse Lane
Red Hook, NY 12571 USA
Phone: 845-758-0400
Fax: 845-758-2634
Email: curran@proceedings.com
Web: www.proceedings.com

 Conference collection

Stress Induced Phenomena and Reliability in 3D Microelectronics

Kyoto, Japan
28–30 May 2012

Editors

Paul S. Ho
University of Texas, Texas, USA

Chao-Kun Hu
IBM T.J. Watson Research Center, New York, USA

Mark Nakamoto
Qualcomm Incorporated, California, USA

Shinichi Ogawa
Advanced Industrial Science and Technology, Ibaraki, Japan

Valeriy Sukharev
Mentor Graphics Corporation, California, USA

Larry Smith
SEMATECH, New York, USA

Ehrenfried Zschech
Fraunhofer Institute for Ceramic Technologies and Systems, Dresden, Germany

All papers have been peer reviewed.

Melville, New York, 2014
AIP Proceedings

Volume 1601

To learn more about AIP Proceedings visit **http://proceedings.aip.org**

Editors

Paul S. Ho
University of Texas at Austin
Microelectronics Research Center
Mechanical Engineering Department
10100 Burnet Road, Bldg. 160
Austin, Texas 78758
USA

E-mail: hops@austin.utexas.edu

Chao-Kun Hu
IBM T.J. Watson Research Center
1101 Kitchawan Road, Route 134
Yorktown Heights, NY 10598
USA

E-mail: haohu@us.ibm.com

Mark Nakamoto
Qualcomm Inc.
5775 Morehouse Drive, R-106E
San Diego, CA 92121-1714
USA

E-mail: nakamoto@qti.qualcomm.com

Shinichi Ogawa
Advanced Industrial Science and Technology (AIST)
West 7A, 16-1 Onogawa, Tsukuba
Ibaraki, 305-8569
Japan

E-mail: ogawa.shinichi@aist.go.jp

Valeriy Sukharev
Mentor Graphics Corporation
46871 Bayside Parkway
Fremont, CA 94538
USA

E-mail: Valeriy_Sukharev@mentor.com

Larry Smith
SEMATECH
3D Interconnect Division
257 Fuller Road, Suite 2200
Albany, NY 12203
USA

E-mail: larry.smith@alum.mit.edu

Ehrenfried Zschech
Fraunhofer Institute for Ceramic Technologies and Systems IKTS
Maria-Reiche-Strasse 2
D-01109 Dresden,
Germany

E-mail: Ehrenfried.Zschech@ikts-md.fraunhofer.de

Authorization to photocopy items for internal or personal use, beyond the free copying permitted under the 1978 U.S. Copyright Law (see statement below), is granted by the AIP Publishing LLC for users registered with the Copyright Clearance Center (CCC) Transactional Reporting Service, provided that the base fee of $30.00 per copy is paid directly to CCC, 222 Rosewood Drive, Danvers, MA 01923, USA: http://www.copyright.com. For those organizations that have been granted a photocopy license by CCC, a separate system of payment has been arranged. The fee code for users of the Transactional Reporting Services is: 978-0-7354-1235-4/14/$30.00

 © 2014 AIP Publishing LLC

No claim is made to original U.S. Government works.

Permission is granted to quote from the AIP Conference Proceedings with the customary acknowledgment of the source. Republication of an article or portions thereof (e.g., extensive excerpts, figures, tables, etc.) in original form or in translation, as well as other types of reuse (e.g., in course packs) require formal permission from AIP Publishing and may be subject to fees. As a courtesy, the author of the original proceedings article should be informed of any request for republication/reuse. Permission may be obtained online using RightsLink. Locate the article online at http://proceedings.aip.org, then simply click on the RightsLink icon/"Permissions/Reprints" link found in the article abstract. You may also address requests to: AIP Publishing Office of Rights and Permissions, Suite 300, 1305 Walt Whitman Road, Melville, NY 11747-4300, USA; Fax: 516-576-2450; Tel.: 516-576-2268; E-mail: rights@aip.org.

ISBN 978-0-7354-1235-4 (Original Print)
ISSN 0094-243X
Printed in the United States of America

AIP Conference Proceedings, Volume 1601
Stress Induced Phenomena and Reliability in 3D Microelectronics

Table of Contents

Preface: Stress-Induced Phenomena and Reliability in 3D Microelectronics
Paul S. Ho, Chao-Kun Hu, Mark Nakamoto, Shinichi Ogawa, Larry Smith, Valeriy Sukharev,
and Ehrenfried Zschech   1

OVERVIEW ON STRESSES AND RELIABILITY IN 3D STRUCTURES
Stress management for 3D through-silicon-via stacking technologies - The next frontier -
Riko Radojcic, Matt Nowak, and Mark Nakamoto   3

**Multi-scale simulation flow and multi-scale materials characterization for stress management
in 3D through-silicon-via integration technologies – Effect of stress on 3D IC interconnect
reliability**
Valeriy Sukharev and Ehrenfried Zschech   18

**Characterization of thermal stresses and plasticity in through-silicon via structures
for three-dimensional integration**
Tengfei Jiang, Suk-Kyu Ryu, Jay Im, Rui Huang, and Paul S. Ho   55

Microstructure, impurity and metal cap effects on Cu electromigration
C.-K. Hu, L. G. Gignac, J. Ohm, C. M. Breslin, E. Huang, G. Bonilla, E. Liniger, R. Rosenberg,
S. Choi, and A. H. Simon   67

Advanced concepts for TDDB reliability in conjunction with 3D stress
Martin Gall, Kong Boon Yeap, and Ehrenfried Zschech   79

ELECTROMIGRATION, STRESS AND RELIABILITY IN 3D STRUCTURES
Electromigration void nucleation and growth analysis using large-scale early failure statistics
M. Hauschildt, M. Gall, C. Hennesthal, G. Talut, O. Aubel, K. B. Yeap, and E. Zschech   89

Modeling of microstructural effects on electromigration failure
H. Ceric, R. L. de Orio, W. Zisser, and S. Selberherr   99

Physics-based simulation of EM and SM in TSV-based 3D IC structures
Armen Kteyan, Valeriy Sukharev, and Ehrenfried Zschech   114

TCAD modeling of stress impact on performance and reliability in 3D IC structures
Xiaopeng Xu and Aditya Karmarkar
128

Design for reliability of BEoL and 3-D TSV structures – A joint effort of FEA and innovative experimental techniques
Jürgen Auersperg, Dietmar Vogel, Ellen Auerswald, Sven Rzepka, and Bernd Michel
138

MULTI-SCALE MATERIALS PARAMETERS, SIMULATION AND CHARACTERIZATION

Thermomechanical characterization and modeling for TSV structures
Tengfei Jiang, Suk-Kyu Ryu, Qiu Zhao, Jay Im, Paul S. Ho, and Rui Huang
148

Assessment of fracture and elastoplastic properties of thin and very thin films
M. Trueba, D. Gonzalez, I. Ocaña, M. R. Elizalde, J. M. Martinez-Esnaola, M. T. Hernandez, M. Haverty, G. Xu, and D. Pantuso
158

3D imaging of semiconductor components by discrete laminography
K. J. Batenburg, W. J. Palenstijn, and J. Sijbers
168

Effects of fluoride residue on thermal stability in Cu/porous low-k interconnects
Y. Kobayashi, S. Ozaki, and T. Nakamura
180

Preface: Stress-Induced Phenomena and Reliability in 3D Microelectronics

This volume of proceedings contains papers from three International Workshops "Stress Management for 3D ICs using Through Silicon Vias – Design for Reliability" held in Santa Clara/CA, USA, on March 17, 2011, in San Francisco/CA, USA, on July 13, 2011, and in Dresden, Germany, on October 12, 2011, as well as from the 12th International Workshop on Stress-Induced Phenomena in Microelectronics held in Kyoto, Japan, on May 28-30, 2012.

The Kyoto Workshop continued the spirit of the stress workshop series to focus on current research relating to stress-induced phenomena in metallization. This workshop was held at a time when continued device scaling has brought significant challenges to development, manufacturing and cost of the Cu/low-k stacks and of future interconnects for the 22 nm CMOS technology node and beyond. The technology challenges have led to the development of the 3D IC integration which provides a potential solution to overcome the wiring limit imposing on interconnect density, performance and power consumption of integrated circuits.

Management of mechanical stress is one of the key enablers for the successful implementation of 3D-integrated circuits using through silicon vias (TSVs). The potential stress-related impact of the 3D integration process on the product reliability must be understood and shared, and designers need a solution for managing stress. The three workshops in San Diego, San Francisco and Dresden provided a forum for discussing stress-related issues in 3D-stacked products with the particular focus on Design-for-Reliability (DFR). The participants of this series of workshops proposed a simulation stress management flow to support a DFR-like solution that would enable design entities to model stress implications on their designs quantitatively. They focused on multi-scale modelling and simulation, multi-scale materials parameters and multi-scale analysis. The experts agreed that the proposed design methodology and the materials properties needed as well as the identified characterization techniques are the right ones.

The papers from these workshops have broadened the scope of the proceedings to include current research on stress management, multi-scale simulation and large-scale failure statistics for 3D and Cu/low k interconnects.

The editors gratefully acknowledge the contribution of time and effort from the members of the Program Committees to make these successful workshops. The Program Committees consist of Reinhold Dauskardt (Stanford University), Alex Dommann (CSEM, Switzerland), Narciso Gambacorti (CEA/LETI, France), Francesca Iacopi (GLOBALFOUNDRIES, US), Kazuhiro Ito (Kyoto University), Young-Chang Joo (National Seoul University), Junichi Koike (Tohoku University),

Klaus-Dieter Lang (Fraunhofer IZM, Germany), Jon Molina (IMDEA Materials, Spain), Tomiji Nakamura (Fujitsu Laboratories, Japan), Tony Oates (TSMC, Taiwan), J. Pyun (Samsung, Korea), Riko Radojcic (Qualcomm, US), Bob Rosenberg (IBM), Larry Smith (SEMATECH), Ralph Spolenak (ETH Zuerich), Olivier Thomas (Aix-Marseille Université, France), King-Ning Tu (UCLA), Kazuyoshi Ueno (Shibaura Institute, Japan), Xiaopeng Xu (Synopsys, US) and Shinji Yokogawa (Renesas Electronics, Japan).

The editors gratefully acknowledge SEMATECH and Fraunhofer for supporting the workshops "Stress Management for 3D ICs using Through Silicon Vias – Design for Reliability". We also thank Jo Ann Smith, Linjun Cao and Tatjana Heller for their assistance in completing the proceedings.

Paul S. Ho, Chao-Kun Hu, Mark Nakamoto, Shinichi Ogawa, Larry Smith, Valeriy Sukharev, and Ehrenfried Zschech

April 2014

Stress Management for 3D Through-Silicon-Via Stacking Technologies
– the Next Frontier -

Riko Radojcic, Matt Nowak, Mark Nakamoto

Qualcomm Inc., San Diego/CA, USA

Abstract. The status of the development of a Design-for-Stress simulation flow that captures the stress effects in packaged 3D-stacked Si products like integrated circuits (ICs) using advanced via-middle Through Si Via technology is outlined. The next set of challenges required to proliferate the methodology and to deploy it for making and dispositioning real Si product decisions are described here. These include the adoption and support of a Process Design Kit (PDK) that includes the relevant material properties, the development of stress simulation methodologies that operate at higher levels of abstraction in a design flow, and the development and adoption of suitable models required to make real product reliability decisions.

INTRODUCTION

It is clear that managing mechanical stress has moved to the forefront of the semiconductor technology, both in terms of achieving the desired device performance through strain engineering techniques in advanced Si technologies and in terms of managing Chip-Package Interactions (CPI) with advanced packaging technologies. Dealing with mechanical stress is vital for successful continuation of technology scaling along the More-Moore axes and is becoming increasingly challenging. Managing mechanical stress factors is even more important for successful deployment of the More-than-Moore type of technologies. The diverse challenges of managing stress/strain characteristics, including both the impacts on device mobility and material integrity, are converging with the Through Silicon Via (TSV) based 3D IC stacking technologies (here referred to as "TSS" for Through Si Stacking technology). Successful dealing with the mechanical stress is thus critical for the deployment of this class of technologies.

The reason and rationale for the need to develop a methodology which would enable users to simulate the effect of mechanical stress in a 3D Si stacked product has been defined before and described in ref 1. A series of industry events were organized in 2010 and 2011 to define a methodology for modeling stress in 3D IC's, and consensus has been reached on (2):

- a DFM-like simulation methodology to model stress characteristics at time zero
- the material properties required to fuel this simulation methodology
- a set of characterization methodologies required to measure these material properties.

These objectives have been accomplished and a commercially supported set of tools and services can now be acquired to model the stress effects in 3D stacked IC products at time zero (3,4,5,8).

In addition, an overall strategy for developing a design-for-stress practice has been defined and agreed on (2). This is pictorially illustrated in the Figure 1 below, and the cross-industry effort now should evolve to phase 2 of the enterprise, and drive adoption of the Design for Manufacturing (DfM) practices, while ramping up a Design for Reliability (DfR) methodology.

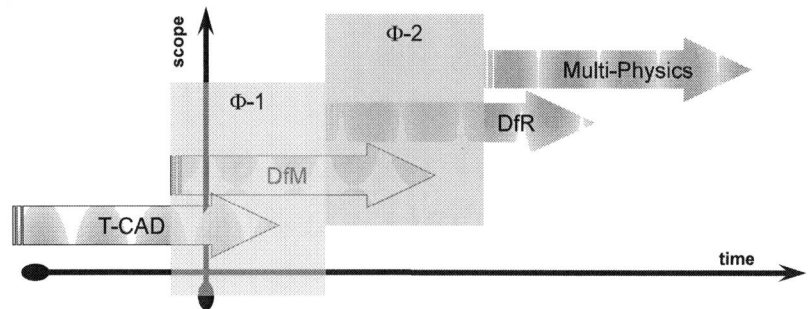

FIGURE 1. Outline of a Strategy for development of Design for Stress Practice

That is, for the purposes of this White Paper, and the cross functional effort, Design for Manufacturability (DfM) is defined as an ability to model stress effects at time zero, after completion of the manufacturing processes, where the sources of the mechanical stresses arise from the intrinsic properties of the materials used, and the characteristics of the manufacturing process itself. Design for Reliability (DfR) is defines as ability to model stress effects through time, with additional externally applied sources of mechanical stresses.

Phase 1 of the enterprise focused on the development and deployment of the DFM-like simulation methodology. This has been accomplished and successfully demonstrated. Phase 2 of the effort is intended to drive the proliferation of this methodology in order to mainstream its support in the supply chain, and to initiate development of DfR methodologies and models.

The intent of this White Paper is to outline the needs and requirements for Phase 2 of the effort.

Note that this White Paper is biased towards the perspective of Fabless/Integrated Fabless Manufacturers (IFM) entities, and that the focus is on managing the mechanical stresses through design, rather than process 'knobs'. This is not to minimize the importance of the process optimization for stress – which is essential for successful deployment of 3D products. It is rather an approach in order to enable fabless design teams to optimize design 'knobs' within the latitude allowed by a given process technology.

MECHANICAL STRESS AND ITS CONSEQUENCES

The Consequences of mechanical stress in semiconductor products, i.e. the principal failure modes associated with the stress driven failure mechanisms, can be segregated into two classes. These are the 'mechanical integrity failure modes' and the 'electrical performance failure modes'. Both are obviously equally deleterious and equally undesirable, as both can cause a failure of a semiconductor device.

Type 1 Consequences: Mechanical Integrity Failure Mode

Examples of mechanical integrity failure modes include cracking, delamination, fracturing, fatigue and other such classical mechanical phenomena. These failure modes typically occur at the interfaces and typically involve passive components of the IC device – e.g. the BEoL wires or the chip-to-package, or package-to-board interconnect wires. These are normally associated with the Chip-Package Interactions (CPI), typically caused by integration of asymmetric materials, with different hardness (e.g. soft ILD glasses and hard Cu pillars) and/or different CTE characteristics (e.g. Si and organic resin based PCB materials), and are usually detected as electrical shorts or opens. Some of these failure modes are illustrated in Figure 2.

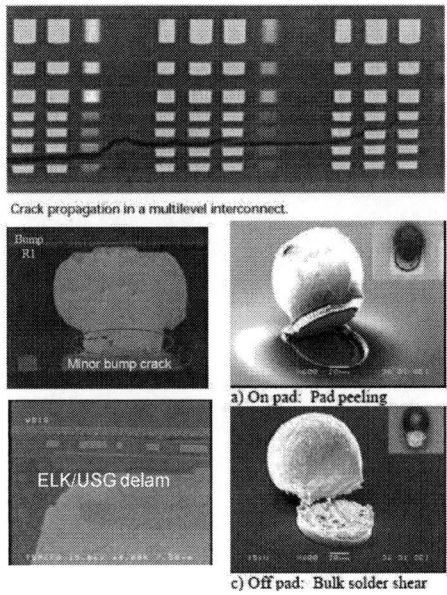

FIGURE 2. Typical examples of CPI issues and "Mechanical Integrity Failure Modes"

Such phenomena are relatively easily detected, and they are normally managed through control of material properties (e.g. amount of Sn in solder, or amount of filler in underfil materials) or through control of process parameters (e.g. temperature excursions or handling jigs). Semiconductor industry has some experience managing this class of failure modes, and whereas the failure analyses and process control are by no means trivial, managing these challenges is relatively well understood and is equally important for the classic 2D products as well as the emerging 3D stacked devices.

From the commercial product point of view, this implies that of course all precautions should be taken to avoid the failure mode, but if the failure mode does take place, the fault diagnostics and correction could be implemented in a reasonable time, resulting in manageable product liabilities.

Type 2 Consequences: Electrical Performance Failure Mode

The second class of possible failures caused by mechanical stress is the class of 'Electrical Performance Failure Modes'. In this case it is the active device – the transistor – that is affected, and the failures are caused by distortions in the crystal lattice of Si, resulting in change in the carrier mobility.

Strain engineering practices used to boost the device performance in modern CMOS technologies through, for example, introduction of Ge doping of Si source/drain regions, is a classic example of constructive use of this mechanism. Figure 3, below, shows the dependence of mobility on stress, as a function of crystal orientation.

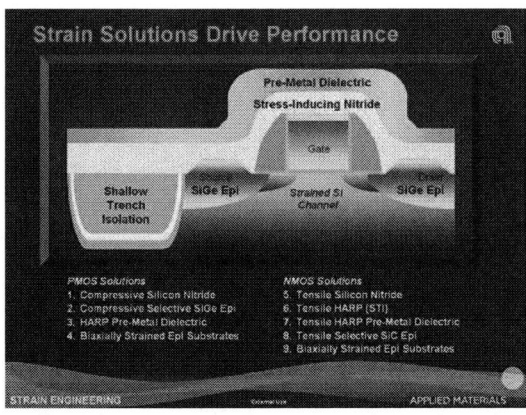

FIGURE 3. (a) Summary of the typical Strain Engineering solutions used to manage the performance of CMOS devices (courtesy Applied Materials)

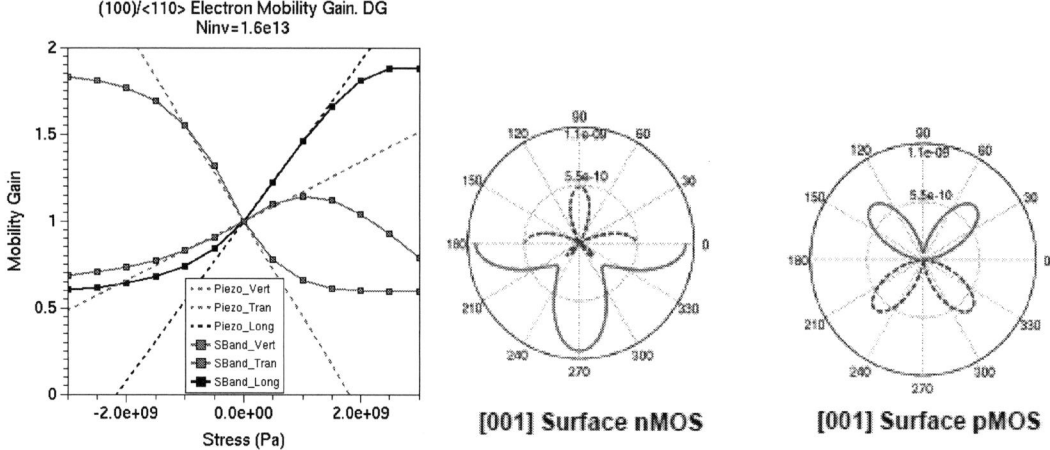

FIGURE 3. (b) Dependence of carrier mobility on strain as a function of crystal orientation and device type (courtesy Xiaopeng Xu, SNPS, ref 5,8)

Managing and controlling the strain in Si devices is extremely complex, depends on device and crystal orientation, device size and type, etc., and mandates use of very restrictive design rules and layout practices. These layout rules are predicated on the conventional 2D product configurations, where the Si substrate used in the device is thick (100's of um) and serves as a solid, stiff reference that minimizes the unintended interactions with other sources of mechanical stress. That is, with the conventional 2D devices, the effect of external stresses is negligible relative to the intended strain engineering stress levels (typically of the order of several GPa) and the effects are included in the models that describe the device characteristics (e.g. BSim models used in SPICE simulations) that comply with the specified layout restrictions.

However, the TSV technology used in 3D stacked devices mandates thinning of the Si substrate down to ~50um, or less, and requires the use of metallization on both sides of the die. In addition, 3D devices obviously use Cu filled TSVs, along with Cu metallizations for both frontside and backside interconnect. The combination of all these factors potentially leads to significant warpage and to external, incremental, stresses driven by packaging or layout interaction, that might be significant relative to the intended strain engineered into the CMOS devices. Incremental stress levels of the order of 100s of MPa are quite feasible, potentially resulting in mobility changes in the order of several or several 10 %. This change potentially results in apparently spurious and local deviation in carrier mobility and shifts in basic device characteristics such as IDsat. If these deviations are not comprehended in device models used in IC design, and/or prohibited by the design rule restrictions, then the transistors do not perform as intended, potentially resulting in timing or levels type of parametric failures, rather than the shorts/opens or functional failures.

The commercial product risks for this type of failure mode are much higher than for the mechanical integrity failure modes for a number or reasons. Since there are no physical manifestations that can be conclusively detected with the usual microscopy and material analyses type of tools, Failure Analyses and Isolation for this failure mode is far more complex – verging on impossible. In addition, it is likely that this type of parametric failures may occur only in some process corners – representing the extremes of normal distribution of process driven variables – such that the failure may appear to be random and not easily reproducible. Since the process variables that interact with this failure mode include not only the variability in the native CMOS process (mobility, Vt, Tox, IDsat, etc..) but also the variability associated with the TSS processes (die or liner oxide thicknesses, material characteristics, etc..), as well as with the layout (device size and neighborhood) and design (device role in a circuit) attributes, the failure diagnostics and correction would take forever. At the product level all this would result in sub-standard and sporadic yield, which normally leads to product commercial failures. Furthermore, this type of parametric failures may be precipitated only by some use conditions, such as for example some system application, and may not be reproducible in other conditions, such as for example the ATE tester, resulting in dissatisfied and angry customers - which may lead in commercial space, to not just a product failure, but a failure of an entire brand. This is a scenario that results in unmanageable product liabilities.

Therefore, it is clear that it is absolutely vital to avoid both types of failure modes. But, avoiding Type 2 failure mode, necessitates a proactive, modeling and simulation based methodology that is baked into the design of 3D products. That is, whereas there is a possible recovery from Type 1 failures, this option is unlikely for Parametric Failures and hence it must be designed out, rather than managed. Ability to model and simulate the Electrical Performance Failure Mode is a requirement for successful deployment of products that leverage the value proposition offered by the 3D TSS technologies.

COPING STRATEGIES FOR STRESS DRIVEN FAILURE MODES

The focus of this White Paper is on managing the stress phenomena through design and design infrastructure, rather than process based controls. There is no doubt that choice of materials, process conditions, etc., are the principal parameters that affect stress-related issues, and that it is the 'process knob' that the industry has used in the past. However, with the emerging 3D technologies, it is believed that the margin for stress related phenomena is reduced, and that the degrees of freedom in the process domain may be limited, thereby forcing more careful control of design parameters.

Design Practices for Control of the Stress Effects

Standard and conventional approaches practiced throughout the industry for managing the effects of mechanical stress and the associated consequences of either type (Mechanical Integrity Failure Mode or Electrical Performance Failure Mode), are to define a series of incremental layout rules that minimize the stress interactions and the consequent deleterious effects. Typical examples of such rules include:

- Rules for Placement of Package or Flip Chip Balls – defined in order to avoid ball shear caused by stress precipitated by difference in CTE between the layers connected by the balls. This type of stress effect is a function of Distance to Neutral Point (DNP), and hence for large chips and/or packages it is not unusual to depopulate the corners, or to ensure that positions with large DNP are occupied by redundant pins (e.g. power or ground pins).
- Rules for Placement and Layout of Features under Flip Chip Balls and/or Cu Pillars – defined typically for technologies that use soft low k ILD. FC Bumps / Cu Pillars are relatively hard materials when compared to porous ILD, and act as concentrators (or transmitters) of stress caused by underfill or mold compound shrinkage. Hence it is not unusual to prohibit placement of sensitive circuits under these pins and/or to dictate specific layout of the BEoL features.
- Keep Out Rules around TSVs – defined to prohibit placement of active devices in the proximity of Cu-filled TSVs. It is well known by now that the Cu-filled TSV exerts stress on the surrounding Si lattice precipitating a complex set of effects on carrier mobility as a function of distance from the TSV – as shown in Figure 4. This rule is typically derived by placing a series of devices at different distances from the TSV and measuring the effect on IDsat, as shown in Fig 4b.

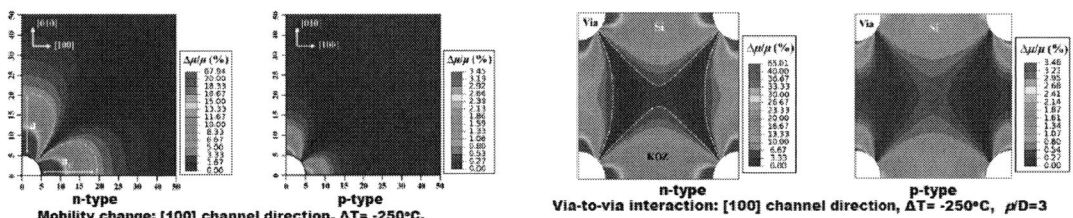

FIGURE 4. (a) Example of a spatial plot showing change in carrier mobility around a Cu filled TSV (courtesy prof Paul Ho, UT, ref 6)

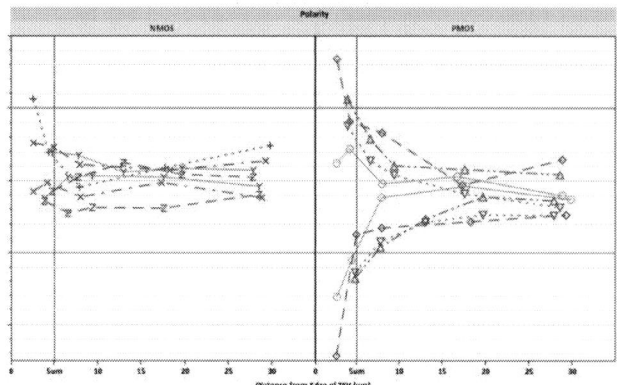

FIGURE 4. (b) Example of the measured IDsat as a function of distance to TSV, measured on a test vehicle

Clearly, such rules are a strong function of the materials used, process technology, manufacturing process conditions, integration schemes, device types and nature, and even application space. They are complex and are intended to shield products from complex multi-dimensional and multi-physics type of phenomena.

Current State of the Art of the Infrastructure for Control of the Stress Effects

The intent of this section of the White Paper is to discuss the various possible mechanisms for developing such rules, and to explore the tradeoffs involved. Specifically a strategy is required to

(a) Define and calibrate stress effect management rules, and
(b) Validate the applicability and effectiveness of such design rules
(c) Ensure compliance with the rules and absence of unintended interactions for any one design.

- **Retroactive Product Correction**
  One strategy is to allow enough time and resources in the product development plan to correct any yield or reliability issues that may come up. If one assumes that each product will behave differently (so that learning based on test chips is not sufficient), or that the learning from one product can be carried to another one (so that the risks to a product are small), then this approach is viable. If the team is lucky and there are no issues – then this is lowest cost and fastest schedule approach that can be implemented with least overhead. It is prudent to insure that, in case a problem does crops up, there is sufficient slack in schedule and resource plan to perform failure analyses, diagnose the problem and correct the issue. As discussed using this strategy for managing stress related phenomena is risky, especially with the Electrical Performance Failure Mode, as failure analysis is (nearly) impossible.

- **Product Like Test Chip**
  An alternative approach could be to define a test vehicle which is expected to be very much like a product and which is designed to include features and configurations which are likely to precipitate any and all yield and reliability issues. The advantage of this approach is that the test vehicle would be presumably run ahead of the product, thereby minimizing the risks to the actual product. In addition, this kind of a test vehicle can be designed to maximize the observability, so that the identification and diagnoses of any stress related failure modes would be easier and faster. The down side of this strategy is the non-productive cost, in terms of schedule and resources, of developing, building and characterizing this test vehicle. Since managing stress-related phenomena is currently mostly based on empirical experiences, rather than some closed form model, exercising a specific layout configuration in order to affirm use of some design rule, using this class of a test vehicle, is quite usual.

- **Technology Test Chip**
  Another approach is to define a set of rules and models and to shelter a product from any issues by ensuring compliance with these rules. The rule set is typically derived from characterization of a technology test chip that includes discrete test structures that are designed specifically for observability and testability of a given stress related failure mechanism. The advantage of this approach is that the observability is maximized and various

mechanisms can be properly characterized and various contributing factors can be separated. The down side is the cost and effort – albeit this type of test vehicle is typically cheaper than product-like test chip. These types of test chips are also typically run on sample and skew lots yielding very limited statistics. Use of technology test chips to characterize given stress related concerns and failure phenomena, especially when ramping up new technologies or material sets, is quite common. The sample data presented in Fig 4b is a typical output from just such a test vehicle.

- **Rules and Corner Models**

Most known failure mechanisms are managed through a set of design and layout rules and/or corner models – as indicated above. These guide the designers to layout a specific product feature around known issues, and to expect given worst case (corner) behavior. A standard practice is to define the rules and corners with a lot of excess margin, in order to avoid problems due to possible interactions and specific weaknesses. The advantage of this practice for managing specific failure mechanisms is that it is relatively simple, usually is based on a single 'black vs. white' kind or a rule, and the design infrastructure is well suited to incorporating rule checkers. Similarly, if corner models are used, they typically define a single, worst case, combination of conditions and expected behavior can be verified through a single simulation or calculation. The limitation of this approach is that it requires a lot of excess margin, and that it is difficult to interpolate or extrapolate from one observed case to another situation.

As indicated, most of the stress-related phenomena are currently managed through a set of incremental design rules and restrictions. Use of layout restrictions to manage the Mechanical Integrity Failure Mode phenomena is intuitively obvious – whether through prescription (e.g. use of a pre-designed and proven FC bumps and the underlying metallization is often mandated) , or prohibition (e.g. placement of circuits in a given relationship to FC bumps or TSVs is often prohibited) . This is the current state of the art for stress management.

With the stress-driven Electrical Performance Failure Mode, the situation is more complicated, and since stress causes a change in device performance, there is an opportunity for optimization. For example, different Keep Out Rules could be applied for different device types – based on their type (nMOST vs. pMOST) size (L and W – especially L) and orientation. Alternatively, there could be an opportunity to pre-characterize given types of circuits and IP blocks as a function of stress, and the use an appropriate model when simulating circuit behavior (this would be analogous to the way that temperature is managed in a normal design flow). It would also be possible to include the performance shifts caused by stress into the overall variability budget and just comprehend these effects in standard device BSim models – resulting in either having different models for 2D and 3D implementation, or carrying excess variability in 2D models. Whereas all these opportunities are possible options, in practice, the complexity that would be created by leveraging them is unacceptable. Hence the standard practice for managing stress driven performance degradation is to use a rule-like one size fits all Keep Out Zone.

- **Physical Models and Simulations**

Some phenomena can be sufficiently well understood that they can be described by physical models, and the behavior of a failure mechanism can be simulated. This for example is the case for the electrical device behavior, and BSim models and SPICE simulations are the fundamental practice for managing timing and voltage levels in modern CMOS products. This methodology allows designers to explore various tradeoffs and to truly optimize a design for a given application.

Currently, stress-related mechanisms and the associated phenomena, however, can barely be simulated even for the simple set of features. Modeling requires intimate knowledge of material properties and process history, and simulation is cumbersome and computationally expensive. Therefore model based approach today cannot be applied at a full chip level, and is used mostly as a T-CAD aide to model the process rather than a design.

In general, all of the above mechanisms are used, and typical stress management strategy involves deploying different aspects at various stages of a technology development cycle. Technology test chips are used to 'calibrate' the design rules and models early in the development cycle, product like test chips are used to 'validate' these before products are deployed, and ultimately products, run in high volume, are used to tune out any outlying issues in volume manufacturing phase and to prove out that the design sign-off procedure is valid. As indicated, models and simulations are rarely used – and if and when they are used, it is for process optimization rather than design release.

For stress driven failure mechanisms and specifically for the mechanisms that drive the Electrical Performance Failure Mode, the standard practices used are immature, and the corresponding infrastructure is not available or ubiquitously deployed even for the standard 2D products. For the emerging 3D products, with their accentuated opportunities for stress interactions, the gaps in the infrastructure are especially acute.

Current State of the Art for Managing Stress Effects in 3D-stacked ICs

Current typical state of the art of the infrastructure for managing stress related issues in 3D technologies is summarized below:

- **Test Chips:** The use of test chips – either the Technology or Product-Like Test Chips – is not unusual, and many studies have been described in the literature (e.g. 2,9,10,11). This is even more so for characterization of the Mechanical Integrity Failure Modes caused by the conventional CPI concerns - highlighted by the migration to the ever softer low k ILD materials, harder Cu Pillar interconnect, and use of ever thinner die. Compared, for example, to the infrastructure (including the test structures, characterization methodologies, model calibration and validation practices, etc…) used to calibrate SPICE models, or printability DFM rules, the stress related test structure and associated practices are immature. The industry is more at the stage of exploring and identifying the right set of tests and test structures for characterizing a given set of stress phenomena, and is not (yet) ready to evolve a standard practices or conventions. Even the basic material characterization techniques and associated benchmark structures, necessary to define the essential parameters such as Young's Modulus or CTE, are not in widespread use or commonly defined across the supply chain. This is especially so for the structures required to characterize the Electrical Performance Failure modes. For example, the criteria for the acceptable amount of mobility shift caused by TSV proximity, measured where and how, and on what kind of a devices, etc., are all undefined.
  Thus, in summary, use of test structures to characterize stress effects is common – but the test structures, and the associated measurement and modeling techniques, have not yet evolved to a level of common convention uniformly practiced across the industry. This is equally true for 2D and 3D technologies.

- **Rules and Models:** The use of design rules and restrictions to mitigate risks associated with the stress related failure modes is not at all unusual and is in common use. As indicated above, the rules used are topically based on empirical experiences, and tend to be one size fits all kind of prohibitions. They are typically calibrated and validated for a simple one-dimensional use case and are not sophisticated enough to explore interactions between multiple layout types or application conditions, etc.. Use of incremental design rules for managing stress effects in 3D technologies, such as Keep Out Zone rules TSV, is common.
  The maturity of the modeling and simulation infrastructure is lower. Classic FEA-type of simulation tools, along with all the limitations associated with application of meshing techniques is used, and it is practiced mostly by highly expert users close to the process development activities. Since, as discussed above, 3D technologies with the extra thin SI substrate may be especially susceptible to complex interactions between different sources of stress (e.g. stress from TSV, from FC Bump, micro-Bump, BEoL, FEoL, etc.) the need for a more sophisticated modeling methodology is especially acute for 3D products. Furthermore, since rule based approach is a poor control mechanism for managing complex interactions in design, deployment of model based methodologies in the design flow (vs TCAD flow) is also quite acute. The efforts described in ref 1 are examples of the most advanced state of the art in the commercial industry.

Consequently, in order to avoid product level yield or reliability failures , which may discredit not only a product, but the entire 3D technology, existing industry practices – and especially the practices necessary to deploy stress simulation and modeling techniques - need to be supplemented and proliferated through the supply chain.

REQUIREMENTS FOR DESIGN-FOR-STRESS PRACTICES

The intent of the rest of this White Paper is to discuss the principal requirements necessary to proliferate the stress management practices in the industry in general, and specifically to enable the entire distributed supply chain to cope with Stress Induced Electrical Performance Failure Modes in 3D stacked products. Again, it is the design perspective that is presented here.

PDK for Stress Modeling and Simulation

It is clear that any meaningful modeling and simulation effort must be based on suitable material properties and corresponding Si characterization techniques. Any model is clearly only as good as the inputs provided, and any simulation result will be only as credible as the relation between reality in Si and model inputs.

- **As Is:** standard practice for managing the process-design interface that the industry has evolved over the years is the use of so called "PDK" (Process-Design Kit). Fundamentally, the overarching objective is to ensure that a given product design is targeted to match the capability of the selected manufacturing process technology. In order to achieve this, the process capabilities and device behavior are fully characterized – typically using a variety of different module and technology integration test chips and in line metrics, etc. . Almost the entire focus of the characterization effort is to describe the physical and electrical attributes of a given process technology. These process characteristics are then captured in a set of files and documents - typically in a format that is sort of standard across the industry that include:

  o Design Rule Manual – basic documents that specifies the fundamental layout rules
  o BSim Model – a fitted physical model that describes the electrical characteristics of all devices
  o DRC/LVS files – files used to verify compliance with design rules
  o LPE files – models and files used to facilitate extraction of parasitic characteristics
  o DFM – models or rules used to guide users towards the technology sweet spot
  o PDK sometimes also include layout constricts, generators, or even some IP elements.

  Note that from the design point of view, the PDK is a full representation of the target manufacturing process and serves as the only interface to the technology. Typically, the PDK and the various files in it, are translated into EDA tool-specific 'technology files' which then fuel the various design steps, through all the levels of design abstraction – from polygon level layout and verification tools, through to place and route, timing and extraction, simulation, floor planning, etc.. Since the designers perceive the process technology through the characteristics of a given design, such as area, performance, power, leakage, etc., it is ultimately the technology files, and the PDK, that drive design optimization procedures, and ultimately the fundamental design-for-x practices. This is illustrated schematically in Fig 5.

- **To Be:** as mentioned above, the entire process characterization practice and PDK implementation is focused on describing the geometric and electrical characteristics of the process technology. As discussed, In order to proliferate stress modeling and simulation necessary to intelligently manage the stress phenomena, various material properties are needed. In order to proliferate the characterization of these properties throughout the supply chain, it would be desirable to leverage the existing paradigms, and, as much as possible, use the existing design tools and flows. Hence, it would be desirable to basically expand the PDK, and the associated infrastructure and practices, to encompass the characteristics required for stress modeling. Material characteristics relevant for stress modeling can then be pulled into the simulation tools through the appropriate 'tech files' and the resulting design optimization learning will ultimately drive a Design-for-Stress culture in the design community. This expanded super PDK should therefore include properties for the variouls materials and films such as

  o Young's Modulus and Poisson ratio
  o Coefficient of Thermal Expansion
  o Residual stresses
  o Glass Transition points.

  The full set of necessary material properties, and the associated characterization techniques are described in (1).

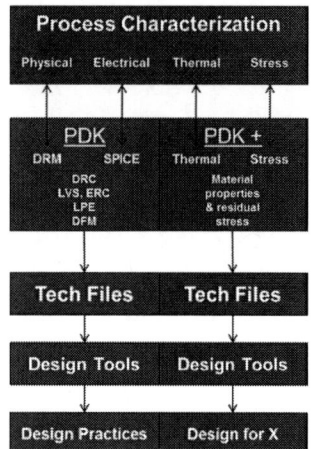

FIGURE 5. Schematic illustration of the PDK and the 'food chain' necessary to drive adoption by the design community. Left side of the chart (in gray) illustrates the existing practices for the geometric and electrical characteristics currently in widespread use. Right hand side (in red) illustrates the proposed expansion necessary for Design-for-Stress practices.

Use of PDK framework would drive some level of standardization and adoption of conventions across the industry. This will then lubricate adoption across the supply chain – on both, the characterization end needed to be supported by the foundries and OSATs, and on the design end necessary for adoption by the EDA companies, and ultimately the design community. This is required in order to drive the proliferation of the stress modeling infrastructure, as illustrated in Figure 5.

Compact Models and Design Abstraction

It is clear that the stress management "knobs" in the process arena are things like choice of materials and process parameters (temperature, pressure, etc.), while 'knobs' in the design arena are parameters such as geometries, sizes, alignments, etc. . And, as mentioned above, it is believed that process 'knobs' alone may not be adequate for future products in general, and for emerging 3D technologies in particular, and that hence process-design co-optimization for stress effects is required.

In order to enable this co-optimization, stress related awareness must be brought into the design flow at appropriate levels of design abstraction. However, it is in the nature of the design methodologies to force a trade-off between accuracy and flexibility. Specifically, normal design flows tend to take design description through successive levels of abstraction – going from high level definition of desired functionality, through definition of connectivity, and down to definition of actual polygon layout and mask works. Intrinsic to the hierarchical nature of the design flow is that the ability to change a design is greatest at the higher levels of abstraction, but that the accuracy of design description is best at the lower levels of abstraction – as illustrated in Fig 6.

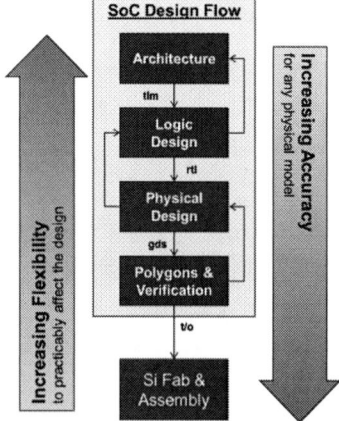

FIGURE 6. (a) Schematic illustration of the hierarchical nature of the Si design flow, the high level hand-off points, and with emphases on trade offs between accuracy and flexibility.

Thus, any physical model, such as for stress related prediction, but also kinetic models, printability models, yield models, etc., are all credible and can be reasonably accurate only at the polygon level of design description. This is intuitively so and intrinsically true. However, performing analyses of a given design for stress (or other physical phenomena) only at this final stage of design flow is unacceptable – because it is too late to do anything about the results. Thus, for example, a modern SoC design , including billion transistors and many billions of polygons, faced with 1000s or 10'000s of 'issues' that require fixing – corresponding to a hit rate of only 1 in a million or better – cannot be practically fixed. Polygon level fixes are usually manually implemented and require a lot of verification. It would take too long. Basically, in practice, a design with this error rate at polygon level, would have to be taken back up to the higher level of abstraction and re-run through the flow, with a different set of constraints.

Conversely, at the higher levels of abstraction, where designers deal with logic level functionality and connectivity, the design can readily be changed and edited – but all physical phenomena can be described only approximately. Thus, for example, the actual timing model can be only an approximation until the wire length and shape is specified exactly, and all the cross-overs and adjacencies are known, etc.. The same would be true for assessment of any stress-related phenomena.

Hence – the nature of the design flow dictates the need for a trade off between accuracy and flexibility – by definition.

- **As Is:** Since this paradoxon is applicable to any physical phenomena, including the electrical behavior, it is not a new challenge in the design domain. Generically, the methodologies that are pursued to address the issue include 'tricks' such as :
  o Use of behavioral models that approximate the anticipated behavior. For example timing engines tend to have a succession of wire load models that approximate wire delays as a function of distance (or other available information) without actually performing an extraction
  o Use of pre-characterized set of constructs and a sort of library of look-up values. For example the timing characteristics of standard cells are pre-characterized and described in .lib models.
  o Use of pro-active rules and constraints that will guide the EDA tools to avoid known bad, or risky, situations. For example Physical Design tools sometimes include forbidden constructs
  o Use of cost-function kind of a model that can be used to make trade off decisions when faced by multi-dimensional problem. For example floor planners tend to rely on cost functions to weigh different possible layout schemes.

  A given physical phenomena – such as timing, or DFM - is then managed through a blend of these kinds of approaches, and fundamentally design constructs are gradually filtered and funneled through the various levels of design hierarchy, such that there is only a relatively small number of issues to be handled at the actual polygon level. This is clearly done at the expense of some excess margin – but is the only practical methodology of the day.

  Currently, clearly there is nothing in the design flow to address stress related phenomena – other than a set of design rules, outlined above, that are used as a sort of a proxy to real physical models.

- **To Be :** Clearly, in order to evolve a true design-for-stress practice, a range of models, algorithms and tools are required to address the stress phenomena at every level of design hierarchy. Based on past experience, methodologies evolve gradually – starting from the lowest level of abstraction (most accurate end to represent physical realities of manufacturing process) and successively bubbling up through the design flow through increasing levels of approximations and behavioral models. Therefore, in order to build a stress awareness into a design flow, and in order to enable designers to manage the stress phenomena (otherwise there is no need to build awareness), stress modeling and simulation methodologies need to evolve. As indicated above, traditionally stress modeling has been used in the TCAD mode in order to optimize process technologies. The current state of the art is moving stress modeling to polygon level of design (2,4,5) – either to build up a library of forbidden constructs, or to use to check the final design for any 'hot spots'. There is currently some initial effort to abstract some stress awareness into Physical Design tools (e.g. (7)). And, eventually, with experience from application of these methods at the lower levels of design, some learning will be built up, and eventually captured in form of constraints to be applied at higher levels of design. This evolution is schematically illustrated in Fig. 6b.

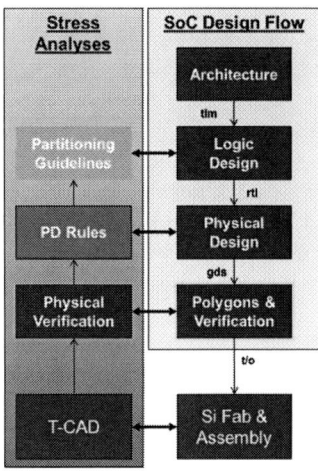

FIGURE 6. (b) A schematic illustration showing the hierarchical nature of design flows and highlighting the need to progressively move stress awareness to the higher levels of abstraction – all in order to ensure that the number of issues to be handled at the polygon level is manageable. The current state of the art is highlighted in red

Use of design hierarchy framework for proliferation of stress awareness through the design community would facilitate adoption of design-for-stress methodologies, because it would enable designers to do something about stress at all levels of design, and thus to optimize products around these (new) considerations.

Reliability Stress Models

As illustrated in Figure 1, hitherto all the effort in the design-for-stress domain has been focused on modeling at time=0, at the end of the fab process line. It is clear that stress effects can also have a significant potential impact on product reliability, and that hence time=t models must also be developed. Whereas the trends that erode the historical excess margins for stress effects are there for all semiconductor products (thin is in, strain engineering is a must on advanced nodes, soft ELK and hard Cu pillars, etc..) , the interactions may be especially acute for the emerging 3D products. Hence ramp up of 3D technology may serve as a convenient compelling event for implementing changes in the traditional reliability practices – that are applicable to both 2D and 3D products.

- **As Is:** The basic Reliability practices have been defined long time ago at the start of the Si technology era, and a set of standard practices have emerged. Fundamentally, it is assumed that the life of a semiconductor device follows the so-called 'bath tub curve' characterized by 3 regions in a failure rate vs. time characteristics, with a rapidly decreasing failure rate during "Infant Mortality" phase, constant failure rate during "Useful Life" and rapidly increasing failure rate during "Wear Out". Typical standard practices that have evolved include :

  o **For Wear Out:** specific failure mechanisms and failure rate distributions are characterized using technology test chip class of vehicles under focused and typically very highly accelerated conditions (Temperature, Voltage, Current Density, etc..). This data is then typically used to define a set of design rules and Safe Operating Area specs which ensure that the wear out does not take place within the targeted useful life. Examples include current density rules for electromigration, max voltage ratings for hot carrier ageing and TDDB, etc. .

  o **For Infant Mortality:** product level screens are applied to weed out the weak parts that are responsible for the Infant Mortality failures. The nature and duration of this screen is a commercial trade off, and can range between something as simple as an extra pass on the ATE testers at elevated voltage or frequency (some ms), through to high voltage/temperature screen (few hours to a full week). Typically these screens are used temporarily while the manufacturing process is tuned so that the infant mortality tail is removed.

  o **For Useful Life:** typically useful life failure rate is demonstrated through a High Temp Operating Life (HTOL) accelerated life test, which can there be modeled to relate the observed (or resolved) failure rate in life test to expected failure rate in filed use. This modeling is focused almost exclusively on temperature driven failure mechanisms, and Arrhenius model with fairly high activation rate is typically used. In addition the component under test is treated more or less as a black box, and the

14

reliability is modeled assuming single uniform condition under life test and a different but also uniform condition in field use (e. g. no attempt to model thermal hot spots, or frequency of operation, or to accumulate wear / relaxation based on operating mode, etc...). In some cases the HTOL test – and the associated models – have been modified to account for failure mechanisms whose behavior is modified by other factors, e. g. humidity (HAST test), Voltage (HVTOL), etc..

o **Everything Else** : for all other eventualities, conditions and failure mechanisms a series of empirical and/or historical 'hammer tests' are performed. Thus, whereas it is known that mechanical stress can drive a number of different failure mechanisms and failure modes, the test that is used to make go/no-go reliability decision for stress related phenomena include thermal cycling or thermal shock, and occasionally a mechanical sequence including drop , vibration and shock tests. Test conditions are fixed, standard tests, performed on fixed sample of the population. If there are no fails – then good – because historical experience says that in-use reliability performance was good on products that passed these tests. Typically there is no attempt to relate the stress conditions to in-use conditions and there is no failure rate model. It is just a pass/fail condition.

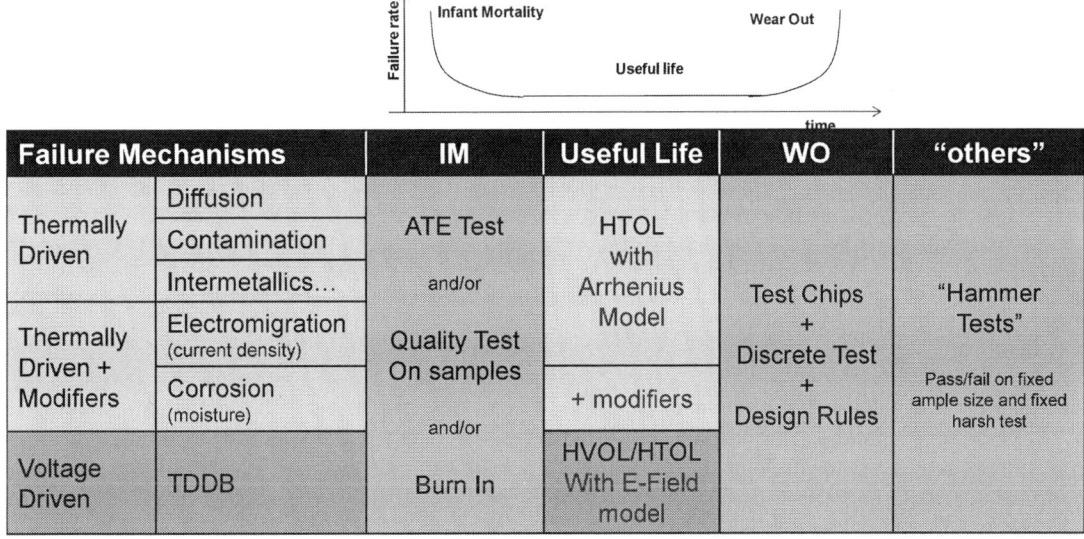

Failure Mechanisms		IM	Useful Life	WO	"others"
Thermally Driven	Diffusion	ATE Test and/or Quality Test On samples and/or Burn In	HTOL with Arrhenius Model	Test Chips + Discrete Test + Design Rules	"Hammer Tests" Pass/fail on fixed ample size and fixed harsh test
Thermally Driven	Contamination				
Thermally Driven	Intermetallics...				
Thermally Driven + Modifiers	Electromigration (current density)		+ modifiers		
Thermally Driven + Modifiers	Corrosion (moisture)				
Voltage Driven	TDDB		HVOL/HTOL With E-Field model		

FIGURE 7. Summary of the current standard as-is reliability practices. As shown, useful life reliability predictions are based almost entirely on the Arrhenius model applicable for thermally driven failure mechanisms.

- **To Be:** The 'hammer test' approach for assessing the impact of stress related phenomena on device reliability has obviously served the industry well up to now. However, history of IC devices is characterized by construction that had a lot of margin for stress driven effects – and especially for the Electrical Performance Failure Mode induced by stress. Thus, typical products were built on thick Si substrates that served as a firm reference plane, used hard glasses for the ILD layers that encased and held the metallization tracks, were single sided construction and used compliant interconnect to the package, etc. . Consequently, the risks of unpredicted design dependent sensitivity for stress driven effects were low. Stress related failures that did occur were the Mechanical Integrity Failure Mode type that is, as described above, relatively easily identified and diagnosed.

The future, however, is likely characterized by devices with less margin for stress-induced failure mechanisms and with new design related sensitivities, especially for the Electrical Performance Failure Mode. Both 2D and the emerging 3D devices are pushing the thickness limits and substrate thicknesses of 100um, 50um and even 30um are now in production – allowing for various new ways of stress relief and strain redistribution. Strain engineering practices used to boost transistor performance makes the devices more sensitive to variability in stress levels. And the use of soft glasses and hard metals enables new models of local concentration of stress levels.

All these trends imply that the industry should be getting ready for the time when products do not pass the standard 1000 thermal cycles pass/fail hammer test, and instead question of how good is good enough will have to be addressed. And, in order to do that, models that relate accelerated test conditions to in-use life conditions for stress driven failure mechanisms shall be absolutely needed. And based on these models, new class of accelerated life tests will need to be developed to qualify products for stress driven failure mechanisms, as illustrated in Fig 8.

Failure Mechanisms		IM	Useful Life	WO	others
Stress Driven	Fatigue ?	?	?	Test Chips + Discrete Test + Design Rules	
	TSV Pop up				
	Others ?				

FIGURE 8. Table that highlights the gaps in the reliability practices for stress driven failure mechanisms. As shown, there are no models that can be used to relate accelerated test conditions to in-use life conditions and that can be used to predict failure rates for stress driven failure mechanisms.

Most of the research and development effort for stress failure phenomena is currently centered on characterizing the intrinsic wear out portion of the reliability characteristics, based on studies using discrete focused test structures and fairly small sample sizes. This work must be supported, expanded and steered in order to develop the needed reliability models.

SUMMARY AND CONCLUSION

This White Paper starts with a summary description of the current status of the development of a Design-for-Stress simulation flow intended to describe the stress effects in packaged 3D stacked Si products using advanced via-middle Through Si Via technology. A simulation flow based on integration of a series of specialized methods to firstly define a set of rules to be used to constrain designs, and secondly to ensure absence of any stress driven process-design interactions, has been demonstrated.

The White Paper focuses on the next steps required to proliferate the methodology in order to ensure development and deployment of the necessary infrastructure throughout the supply chain. The requirements for this proliferation are discussed and three principal regions for future activity are highlighted. The challenges include:

(a) adoption and support of an expanded PDK-like approach to include the relevant material properties and other characteristics of manufacturing process needed to model stress effects in design arena,
(b) development of stress simulation methodologies that operate at higher levels of abstraction in a hierarchical design flow required to enable designers to optimize the design for stress driven phenomena, and
(c) development and adoption of suitable reliability models required to relate accelerated test conditions to in use life conditions for stress-driven failure mechanisms.

The paper discusses the reasons for these requirements and the tradeoffs necessary for their adoption, with emphases on the design activities and opportunities.

REFERENCES

1. R.Radojcic, M.Nowak, M.Nakamoto, "TechTuning: Stress Management Methodology for 3D TSV Technologies", AIP Conference Proceedings, Published July 2011; ISBN 978-0-7354-0938-5
2. E.Zschech, R.Radojcic, V.Sukharev, L.Smith, "Stress Management for 3D ICs using TSV", AIP Conference Proceedings, Published July 2011; ISBN 978-0-7354-0938-5

3. K.Dhandapani, M.Nakamoto, W.Zhao, A.Syed, W.Lin, R.Radojcic, A Methodology for Chip - Package Interaction (CPI) Modeling in 3D IC Structures, Proceedings IMAPS 8th Device Packaging Conference Scottsdale, AZ, March 5-8, 2012

4. X.Xu, A.Karmarkar, "TCAD Modeling and Stress Impact on Performance and Reliability in 3D IC Structures", Proceedings 12th International Workshop on Stress Induced Phenomena, Kyoto, Japan, May 28-30, 2012

5. V.Sukharev, et al., "Multi-scale Simulation Methodology for Stress Assessment in 3D IC", Journal of Electronic Testing, V 28 N1, 02, 2012

6. PS. Ho., et al., "Reliability Challenges for 3D Interconnects: A Material and Processing Perspective", International Stress Workshop, Kyoto, May 28, 2012

7. Jung, J.Mitra, DZ. Pan, SK Lim, "TSV Stress-aware Full-Chip Mechanical Reliability Analysis and Optimization for 3D IC", DAC '11, Jun 05-10 2011, San Diego, California, USA

8. A.Karmarkar, X.Xu, S.Ramaswami, J.Dukovic, K. Sapre, A. Bhatnagar, "Material, Process and Geometry Effects on Through-Silicon Via Reliability and Isolation", Mater. Res. Soc. Symp. Proc. Vol. 1249, 2010

9. GS. Leatherman, J. Xu, J. Hicks, B. Kilic, D. Pantuso, "Die-Package Stress Interaction Impact on Transistor Performance", Proc International Reliability Physics Symposium, 15 Apr - 19 Apr., 2012 Garden Grove, CA, USA, IEEE

10. V. Cherman, J. De Messemaeker, K. Croes, B. Dimcic, G. Van der Plas, I. De Wolf, G. Beyer, B. Swinnen, E. Beyne, "Impact of Through Silicon Vias on Front-End-of-Line Performance After Thermal Cycling and Thermal Storage" , Proc International Reliability Physics Symposium, 15 Apr - 19 Apr., 2012 Garden Grove, CA, USA , IEEE

11. M.Nakamoto, R.Radojcic, W. Zhao ; V.K. Dasarapu, A.P Karmarkar, X. Xu, "Simulation Methodology and Flow Integration for 3D IC Stress Management", Proc. Custom Integrated Circuits Confernce, CICC-2010 IEEE

MULTI-SCALE SIMULATION FLOW AND MULTI-SCALE MATERIALS CHARACTERIZATION FOR STRESS MANAGEMENT IN 3D THROUGH-SILICON-VIA INTEGRATION TECHNOLOGIES – EFFECT OF STRESS ON 3D IC INTERCONNECT RELIABILITY

Valeriy Sukharev, Mentor Graphics Corporation, Fremont, CA 94538, USA
Ehrenfried Zschech, Fraunhofer Institute for Ceramic Technologies and Systems, D-01277 Dresden, Germany

The paper addresses the growing need in a simulation-based design verification flow capable to analyze any design of 3D IC stacks and to determine across-layers implications in 3D IC reliability caused by through-silicon-via (TSV) and chip-package interaction (CPI) induced mechanical stresses. The limited characterization/measurement capabilities of 3D IC stacks and a strict "good die" requirement make this type of analysis really critical for the achievement of an acceptable level of functional and parametric yield and reliability. The paper focuses on the development of a design-for-manufacturability (DFM) type of methodology for managing mechanical stresses during a sequence of designs of 3D TSV-based dies, stacks and packages. A set of physics-based compact models for a multi-scale simulation, to assess the mechanical stress across the dies stacked and packaged with the 3D TSV technology, is proposed. As an example the effect of CPI/TSV induced stresses on stress migration (SM) and electromigration (EM) in the back-end-of–line (BEoL) and backside-redistribution-layer (BRDL) interconnect lines is considered. A strategy for a simulation feeding data generation and a respective materials characterization approach are proposed, with the goal to generate a database for multi-scale material parameters of wafer-level and package-level structures. A calibration technique based on fitting the simulation results to measured stress components and electrical characteristics of the test-chip devices is discussed.

Keywords: Simulation, FEA, Compact model, Materials characterization

1. INTRODUCTION

It is a common understanding that the motivation for 3D IC integration is a mixture of economic and technical requirements, summarized within the term "More than Moore" [1]. 3D IC stacking technologies (including 2.5D interposer-based approaches), applying thinned wafer/through silicon via (TSV) structures, are novel solutions that result in reduced floor space, higher bandwidth and reduced energy consumption. Finally, one of the current challenges is to meet the costs targets. The processing of high-density TSV structures through thinned dies and subsequent 3D stacking is a promising technological alternative to the traditional 2D lithography/etch scaling. However, several issues have to be addressed to guarantee the needed product performance and reliability. In addition to the design-for-manufacturing (DFM)-like approach to addressing the stress related implications in 3D IC performance, which was proposed and extensively discussed in [2], a similar approach to addressing the stress-induced implications in 3D IC reliability has to be developed. The design-for-reliability (DFR)-like methodology, developed on the basis of multi-scale simulation, is discussed in this White Paper. In both cases, the stress-induced and stress-enhanced effects are playing an essential role.

The main element of the extension of the standard reliability tests to 3D IC structures, containing stacks of two or more dies, is the design and fabrication of appropriate test structures to evaluate the reliability of 3D TSV systems. In the case of the true 3D IC TSV technology, the major reliability concerns are caused by the TSV and micro-bump formations as well as by defects generated during backside processing [3]. Reliability issues associated with TSV formation are usually related to the quality of the film stack deposited after etching the TSV structure, including dielectric liner, barrier and metal seed layer. Higher aspect ratio TSVs require more conformal deposition processes for these thin films to guarantee homogeneous thin film formation on the sidewall of the via and the required step coverage. Chemical vapor deposition (CVD) and atomic layer deposition (ALD) processes are advantageous to meet these target requirements. Subsequently, the metal fill process, usually electrochemical deposition (ECD) has to be controlled in order to prevent the formation of voids or seams in TSVs. Both incomplete filling on the sidewall and voids in the bulk can cause reliability issues. The adhesion between silicon die and metal TSV, including the liner/barrier films, is another critical parameter from the reliability point of view. Backside processing for 3D TSV integration includes thinning, etch-back, passivation and the formation of redistribution layers. Most of these process steps are unique to 3D IC stacking, and the associated reliability implications have to be understood.

3D-Reliability Domains

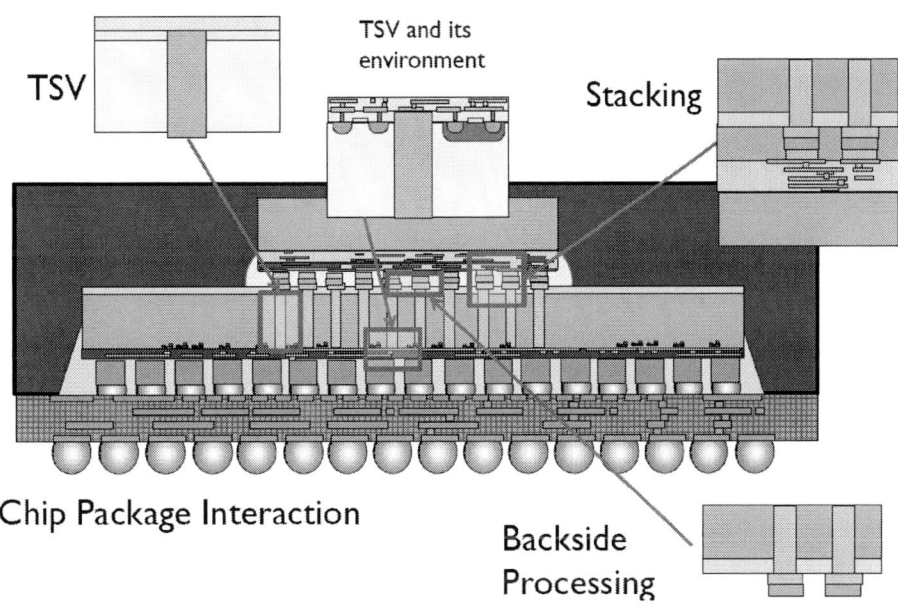

Fig. 1. Thermal-mechanical issues in 3D-DRAM structures [5].

Thermo-mechanical reliability is one of the key concerns for the adoption of the 3D IC technology. Indeed, the presence of extremely thin dies with thicknesses below 50 μm, mounted on arrays of solder bumps, the presence of TSVs, i. e. large metal structures that are several tens or hundreds times larger than typical structures in BEoL stacks, plus operation conditions which are characterized by temperatures and temperature gradients essentially higher and steeper compared to conventional planar chip architectures inevitably generate mechanical stresses of values that significantly exceed the stresses in planar 2D dies. Mechanical stress and crystal defects are introduced into thinned dies not only during wafer or die thinning, but also during wafer bonding using fine-pitch, high-density micro-bumps and curing. Furthermore, metallic TSVs and micro-bumps generate strain/stress in the thinned silicon die due to the difference of the coefficients of thermal expansion (CTE) between Si and metal. Micro-Raman spectroscopy data reveal that a local tensile strain of 1.8 GPa was induced by 4x4 μm^2 square sized Si micro-bumps in 10 μm thick LSI wafers after bonding and curing. It was noticed that this locally induced stress caused more than 10% change in the drive current of a p-MOS transistor [3]. CuSn micro-bumps generate stress in the near-surface region of the Si die, which spreads out deeper for larger bump size and wider for smaller bump pitch. As shown in [4], up to 40%

change in I_{on} current can be caused by micro-bump impacts. A good example of the "inventory" of thermal-mechanical issues expected in 3D IC integration is shown in Fig. 1 [5].

A series of the three consecutive workshops organized by SEMATECH and Fraunhofer IZFP Dresden (now Fraunhofer IKTS Dresden) to address stress management for reliable 3D IC stacks were held in the year 2011. The focus of these workshops was on discussions and considerations of stress-driven failure mechanisms in the 3D IC stacks (Santa Clara/CA, March 17, 2011), product level considerations for dealing with stress-driven reliability mechanisms of the via-middle TSV 3D stacking technology (San Francisco/CA, July 14, 2011), and reliability-limiting degradation kinetics in TSV-based 3D IC stacks (Dresden, Germany, October 12, 2011). Workshop participants have generated a set of tables summarizing the implications of mechanical stress in 3D IC technologies on reliability. These tables 1-5 have put together new reliability-limiting effects that have been arisen in these technologies, failure mechanisms associated with new elements (structure features), prospective tests and test regimes for failure monitoring, detection and characterization, and different approaches for these failures modeling.

Table 1: New elements introduced by the 3D IC technologies in comparison to the planar 2D technology

ELEMENT	Failure mechanisms	Test	Test regimes	Modeling approaches
Thin die	Warpage-induced stress effect on device performance, TSV/Si delamination and silicon cracking, misalignment, deterioration of of BEoL reliability (TDDB) by Cu diffusion from the back side	Ultrasonic microscopy; chain structures; ring oscillators (RO); test for leakage; capacitance measurements, capacitance-time method (C-t)	Temperature-accelerated (T-accelerated) (Constant T anneal for diffusion-driven failures and T-cycles for CTE-mismatch driven failures)	FEA/compact modeling
TSV	Si cracks, BEoL cracks/ delaminations, TSV-stress effects device performance, EM/SM	RO, Test for leakage, capacitance measurements, resistance	T&I (current) accelerated	FEA, compact modeling
Chip-to-chip (C2C) connection/underfill (UF)	Stress generated by chip stacking effects device performance, delamination / BEoL crakcs.	Ultrasonic microscopy, chain structures, RO, Test for leakage, capacitance	T-accelerated	FEA

		measurements			
Die backside redistribution layer (BRDL)	Stress-induced cracking/delamination, Cu contamination on the back surface of the thinned wafer/stress-enhanced Cu diffusion	Ultrasonic microscopy, chain structures, RO, Test for leakage, capacitance measurements, C-t	T-accelerated	FEA, CAD	
Micro-bumps/UBM	Cracking, EM/SM, electrical resistance	Wheatstone bridge + XCT, resistance	T&I-accelerated	FEA/compact modeling	
Large volume of Cu	Cu contamination on the backside surface of the thinned die/stress-enhanced diffusion	Bending Beam, C-t	Use conditions	FEA	
Large thermal gradients	Chip performance, BEoL cracking	Ultra sound spectroscopy, chain structures, RO, Test for leakage, capacitance measurements	Use conditions	FEA/compact modeling	

Table 2: Implications of mechanical stress in 3D IC technologies on reliability: **TSV**

Failure Mechanisms	Driving forces	Outcome	Test	Test regimes	Modeling approaches
1. TSV-induced stress effect on device performance	CTE mismatch	Device performance degradation	RO, device in the vicinity of TSV arrays	Use conditions	FEA, compact modeling
2. TSV effect on BEoL	CTE mismatch	Interface delamination (adhesive failure), ILD cracking (cohesive failure)	chain structures (open/shorts detection)	T-accelerated	FEA

3. Electromigration and stress migration	TSV voids	Resistance increase	Wheatstone Bridge + X-ray microscopy/ tomography, Kelvin/chain structures	T & I– Accelerated	FEA, compact modeling
4. Barrier/oxide liner integrity	CTE mismatch; deposition, interface conditions	Signal propagation	Test for leakage	T-accelerated	FEA
5. Stress-induced corrosion	"Out-of-spec" stress effecting migration/ reaction rates	Metal voiding, delamination	Combined corrosion + stress test	T-accelerated	
6. Interfacial failures: Cu/barrier, barrier/oxide	CTE	BEoL metal/ULK and TSV metal/Si delamination/ cracking	Capacitance measurements; continuity structures (shorts & opens)	T-accelerated	
7. TSV pop-out	CTE mismatch	TSV/Si interface delamination and cracking, TSV bottom voids	nano-XCT	T-accelerated	FEA
8. Copper/TSV CMP process	Interaction with W contacts	Contacts failure	Resistance measurements	Use conditions	CMP models
9. Cu contamination on the backside surface of the thinned die, stress-enhanced diffusion	Cu atoms originated from Cu TSV	Deterioration of device characteristics	C-t	Use conditions	

Table 3: Implications of mechanical stress in 3D IC technologies on reliability: **Micro-bumps**

Failure Mechanisms	Driving forces	Outcome	Test	Test regimes	Modeling approaches

1. Effect of bump proximity on transistor; role of underfill (UF)	CTE mismatch	Performance degradation	RO, device in the vicinity of TSV arrays	Use conditions	FEA, compact modeling
2. IMC: formation of intermetallic phases	Metal-physical processes	Voiding, cracking, EM, power, electrical resistance	Energy-disperive X-ray (EDX) spectroscopy, Auger electron spectroscopy (AES); cross-sectional SEM; Bump shear test and tensile chip pull test; resistance measurements	Use conditions	
3. W2W or C2C joining process	Warpage-induced stress; CTE mismatch, UF/bump-induced pressure	Device performance, IC delamination/cracking, misalignment	Ultra sound spectroscopy, chain structures, RO, Test for leakage, capacitance measurements	T-accelerated	FEA
4. Under bump metallization (UBM)	Metal-physical processes, stress	Voiding, cracking, EM, electrical resistance	Ultrasonic microscopy, X-ray microscopy/ tomography, chain structures (open/shorts detection); Wheatstone Bridge + X-ray; resistance measurements		
5. Cu contamination	Cu atoms originated	Deterioration of device characteristics	C-t	Use conditions	

Failure Mechanisms	Driving forces	Outcome	Test	Test regime	Modeling approaches
on the backside surface of the thinned die/stress-enhanced diffusion	from micro-bumps				
6. Electromigration and stress migration in on-chip interconnects and TSVs	TSV voids	Resistance increase	Wheatstone Bridge + X-ray microscopy/ tomography	T & I– Accelera-ted	FEA, compact modeling

Table 4: Implications of mechanical stress in 3D IC technologies on reliability: **BRDL**

Failure Mechanisms	Driving forces	Outcome	Test	Test regime	Modeling approaches
1. Interfacial RDL/BCB-polymer delamination	CTE	Delamination	Capacitance measurements	T-accelerated	
2. TSV effect on RDL	CTE mismatch	Interface delamination	Ultrasonic microscopy, chain structures (open/shorts detection)	Temperature T-accelerated	FEA
3. Cu contamination on the backside surface of the thinned die/stress-enhanced diffusion	Cu atoms originated from Cu TSV	Deterioration of device characteristics	C-t	Use conditions	

Table 5: Implications of mechanical stress in 3D IC technologies on reliability: **Thin die**

Failure Mechanisms	Driving forces	Outcome	Test	Test regime	Modeling approaches

1. W2W or C2C joining process	Warpage-induced stress; CTE mismatch, UF/bump-induced pressure	Device performance, BEoL cracking	Ultrasonic microscopy, chain structures, RO, Test for leakage, capacitance measurements	T-accelerated	FEA
2. Cu contamination on the backside surface of the thinned wafer/stress-enhanced diffusion	Cu atoms originated from Cu TSV/micro-bumps; wafer or die thinning can remove intrinsic and extrinsic gettering regions (IG & EG)	Deterioration of device characteristics due to reduction of the minority carrier lifetime by impurity levels generated by Cu-atoms	C-t	Use conditions	

Tables 1-5 demonstrate that modeling and simulation can be used to numerically analyze the thermo-mechanical stresses which depend on the materials and materials combinations, on processes and process parameters as well as on the designs of the 3D IC structures. Modeling and simulation can be particularly useful for the design of test-structures and, upon the model calibration is done, for the prediction of reliability-limiting stress-induced degradation and failures in a variety of geometries and local layout configurations.

Modeling and simulation techniques could and should be applied to address some of the 3D stack-related reliability issues, for example:

- Effect of wafer or die thinning (warpage-induced stress) on TSV/Si interface reliability (cracking, delamination, etc)
- Effect of chip-package interaction (CPI) (CTE mismatch, non-uniform solder bump solidification, temperature-induced effects like intermetallic phase formation, etc) on the BEoL, BRDL reliability (interfacial and cohesive cracking, etc)
- EM/SM related failures for different integration schemes for TSVs, BRDL and BEoL interconnect structures
- Effect of Cu TSV architecture, material texture and interfacial adhesion on EM/SM failures.

The preliminary classification of the simulation methodologies and techniques, which depends on the size and physics of the analyzed problem, could be proposed:

FEA-based simulations with the sub-modeling technique:
- TSV-induced stress effect on cohesive and adhesive failure (delamination, cracking)
- TSV-induced stress effect on EM/SM in BEoL and BRDL interconnects
- CPI-induced stress effect on BEoL mechanical behavior (adhesive and cohesive failures)

CAD-based simulations:
- Current densities through BEoL interconnects

- Temperature distributions for BEoL, BRDL and TSV interconnects

TCAD and process simulations:
- Process-dependent feature-scale simulations

Compact model-based simulations:
- Effective mechanical properties of BEoL and BRDL interconnects
- Hot-spot check of EM/SM-induced failures in BEoL and BRDL interconnects-.
- Interfacial and cohesive failures in BEoL and BRDL interconnects.

This list will be extended upon development of new robust simulation models for several reliability phenomena. In the following sections of this White Paper, we will demonstrate some results which have been achieved already by implementing simulation techniques to address the number of reliability issues specific for 3D IC technologies. We will outline the directions of the development of a DFR type methodology for managing mechanical stresses during a sequence of designs of 3D TSV-based dies, stacks and packages. A strategy for a simulation feeding data generation and for a respective materials characterization approach will be proposed, with the goal to generate a database for multi-scale material parameters of wafer-level and package-level structures. A calibration technique based on fitting the simulation results to measured stress components and electrical characteristics of the test-chip devices will be discussed.

2. ADDRESSING THE IMPACT OF STRESS ON 3D IC RELIABILITY BY MEANS OF SIMULATION: CASE STUDIES

A noticeable activity in the simulation of the impact of stress on different aspects of reliability of 3D IC integrated structures has been demonstrated recently. While the majority of papers is still biased toward the estimation of the effect of TSV-generated stress on the carrier mobility in transistors, a significant number of publications is dealing with the thermo-mechanically effected reliability of TSV structures, for example, the nucleation and growth of cracks within a material (cohesive failure) and along the TSV interfaces (adhesive failure) [6-10]. It was demonstrated that interactions between the 3D TSV-induced stresses and interconnect cracks have a significantly adverse impact on the reliability of 3D IC integration structures [10]. Simulation results of the impact of TSVs on the stress distribution in the BEoL interconnect structures have been published in the number of papers, see for example [11-13].

An additional concern associated with the presence of TSVs in a thin die is related to the SM and EM in the BEoL/BRDL interconnect metal segments located in the TSV proximity as well as a possible SM and EM degradation inside the TSV. A major argument is the drastic difference in sizes between TSVs and BRDL/BEoL metal lines and vias. This size difference can generate large stress gradients inside BEoL interconnects during thermal anneal and/or thermal cycling and results the stress-induced voiding. The mentioned stress gradients depend on the TSV type, its architecture and the metal fill process, which can generate different microstructures [14]. Several papers describe results of the simulations of EM in TSVs and neighboring interconnects [13-18]. FEA-based simulations were performed for several types of TSV architectures (via-middle and via-last) which are characterizing by different metal microstructures, and particularly by different grain size distributions [18].

In the next paragraphs, the above-mentioned simulations performed to understand and to predict differences in the reliability of conventional planer chips and their 3D stacked counterparts will be discussed and summarized.

2.1. Thermo-mechanical analysis of Si/TSV structures and the role of TSV-induced stress for interfacial and cohesive cracking

Tanaka, [19], was one of the first who recognized that the large CTE mismatch between the copper TSV and the silicon die is very important for 3D IC reliability, and, hence, the effect of thermomechanical stress has to be taken into consideration. He considered a number of Si/Cu TSV reliability issues including thermal fatigue of copper, delamination at the TSV interface, and silicon die cracking due to thermal stress. These effects can cause catastrophic failures like electrical opens and shorts caused by copper migration. FEA-based simulations were performed to study the effect of thermal stress on interfacial delamination, the damage of TSV caused by loading during the stacking process, and the effect of TSV on the thermal fatigue lifetime of chip-to-chip interconnects. It was concluded that TSVs with smaller diameters have a lower potential for thermal fatigue than the larger ones. Performed simulations and mechanical strength tests demonstrated that copper TSVs were strong enough under the typical bonding conditions. The shear stress between the Cu TSV and the silicon die was large at the edge of the bump at the die interface, but was decreasing rapidly inside the silicon die interior. It was shown that the reliability of the gold bump-based chip-to-chip interconnect largely depended on the bump height and the properties of the underfill material. This result demonstrated the role of the thermal mismatch between the bump and underfill materials responsible for the plastic deformation in the flip-chip (C4) and micro-bumps (u-bumps).

Ryu reported the existence of two modes of interfacial delamination for fully filled TSV structures [9, 10]. To assess the thermo-mechanical reliability of TSV structures, the driving forces for both cohesive and interfacial crack growths were calculated based on fracture mechanics [9]. Fig. 2 shows schematically the interfacial delamination of a TSV under cooling and heating conditions. In both cases, the interfacial crack was assumed to grow axisymmetrically from one surface toward the other surface. In the case of cooling ramp ($\Delta T < 0$), both the shear stress and the tensile radial stress have contributed to the driving force for interfacial delamination (mode I). In the case of heating temperature ramp ($\Delta T > 0$), however, the radial stress was compressive, and consequently, it did not contribute to the driving force for interfacial delamination. The interfacial crack was characterized by a pure shearing mode (mode II). Based on the derived analytical solution for the steady-state energy release rate, several predictions were made. It was found that the decrease in the TSV diameter reduced the stress, and therefore, it could help to avoid delamination. The driving force for delamination could be reduced by either using TSV materials with smaller thermal expansion mismatch to Si and/or by reducing the thermal load. It was found also that the reliability of Cu TSV/Si die interfaces was improved for thinner dies.

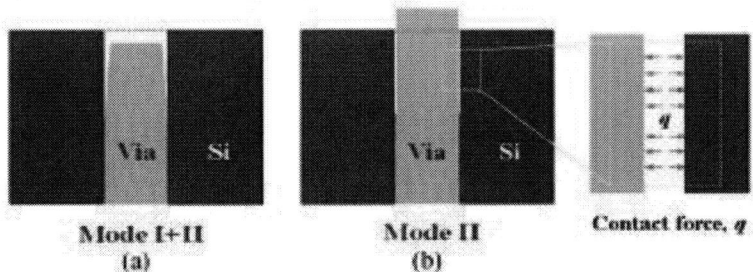

Fig. 2. Interfacial delamination: (a) Cooling ($\Delta T < 0$), (b) Heating ($\Delta T > 0$) [9].

Two-dimensional thermo-mechanical finite-element models were built in [8] in order to analyze the stress/strain distribution in TSV structures. The models confirmed that the large stress gradients and plastic deformation were developed near the corner of electroplated Cu pads. The stresses, simulated with FEA models, were compared with XRD experimental data. An additional fracture mechanics analysis revealed that Cu/SiO$_2$ interfacial cracks and SiO$_2$ cohesive cracks likely were initiated and propagated at corner locations (Fig. 3).

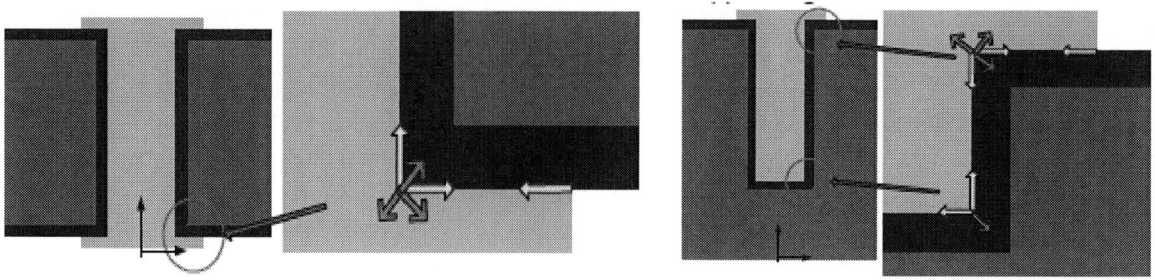

Fig. 3. Geometries of analyzed through-via (left) and blind-via (right) structures. Arrows show the predefined cohesive (green) and interfacial (white) cracks built in the models [8].

It was shown that high a stress concentration, that exist at Cu pad corners of through-vias and blind-vias, caused a plastic deformation of Cu. The fracture analysis not only confirmed those critical locations, but also showed that the failure mechanisms was associated with interfacial delamination at the Cu/SiO$_2$ interface as well as with cohesive cracking of dielectric layers. Furthermore, the FEA-based simulation demonstrated that the energy release rate was increased during interfacial crack growth, resulting in unstable crack growth at the critical corner locations. The energy release rate was increasing with TSV diameter and aspect ratio, which is in agreement with Tanaka's and Ryu's predictions [9, 19].

All results described above were obtained for a single TSV embedded into a thin silicon die (interposer). The accumulated effect of all TSVs resulting in a global mismatch between the interposer and the mounted die, that is responsible for the fatigue loading on the solder joints, was considered in [6]. A rigorous analysis of nonlinear stress/strain caused by the local and global mismatches was presented. The local stress/strain analysis considered the effect of the TSV aspect ratio as well the effect of the BRDL. The global model was used to quantify the lifetime (via the density of creep strain energy per cycle) of the micro-bumps (solder joints) that interconnect the TSVs in the interposer and the on-die interconnects. The performed simulation analysis demonstrated that for all considered TSVs, characterized by aspect ratios larger than 5, there is little dependence of stress and strain on the aspect ratio. It was also shown that for the modeled perfect TSV structures failures at the TSV interfaces (due to the local CTE mismatch between Si, SiO$_2$, and Cu) are unlikely since the strains in these elements are not large enough to cause them.

In relation to the discussed results it should be mentioned that a high amount of copper containing in the interposer increases its effective CTE. Hence, the global thermal mismatch between the Si die and the Si/Cu TSV interposer could be very large, and the micro-bumps connecting them can be subjects for extremely high stresses, especially for the bumps located at the periphery of the stuck. This result strongly emphasizes the necessity of multiple length scales that have to be considered in the thermo-mechanical modeling of 3D IC systems. It is easy to demonstrate that those different elongations or contractions of the dies to be stacked and the layers of underfill with solder joints can result in the complex warpages of individual dies, introducing additional stresses into all structures (Fig. 4). These additional stresses can change, for example, the conditions for crack initiation and its growth behavior.

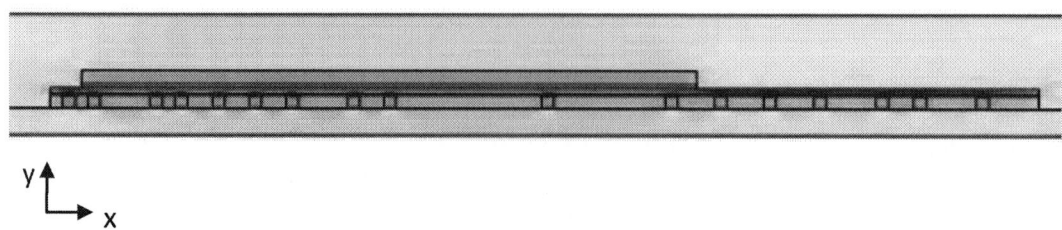

y

x

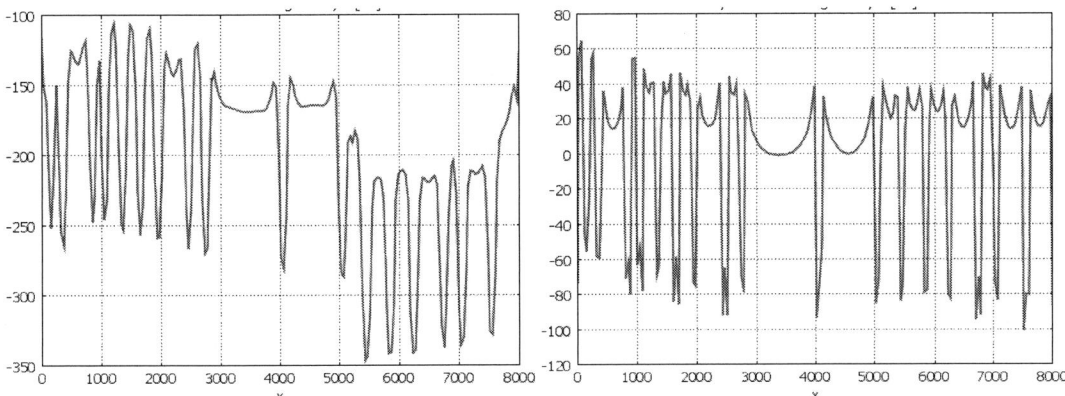

Fig. 4. Simulation setup and distributions of lateral (left) and vertical (right) stresses inside the bottom tier (stresses in MPa and distances in μm).

All papers cited above were dealing with complex problems related to stress-induced reliability of TSV structures in interposer applications. An even more complicated situation is expected if a TSV is crossing the active silicon of, for example, a logic die, which occurs in the case of true 3D TSV-stacked ICs. In this case, new effects should be considered in addition to the already considered reliability impacts caused by thermo-mechanical stress. These effects are related to the impact of TSV-induced stress on BEoL and BRDL interconnects. The interactions of these stresses with preexisting defects can lead to crack nucleation and growth, and compromise the structure reliability. Furthermore, the material choice that reduces the silicon stress for less impact on device performance may increase stresses in other regions where reliability is a concern. For example, the introduction of a wide enough low-k guard-ring at the end of a TSV (near the Si/BEoL interface) can reduce the strain propagation into the nearby transistors, but it makes it easier to generate delamination at the triple point: TSV/Si/BEoL. The most comprehensive study of the interactions between thermo-mechanical stress generated by TSV and BEoL interconnect structures so far was reported in [11]. FEM-based global modeling of TSV structures and the corresponding submodeling for resolving interconnect metal lines (Fig. 5) allowed studying these effects and their dependence on various integration configurations. The stress distributions generated in several BEoL interconnect structures during the thermal ramp were simulated for the cases when copper or tungsten TSVs were presented in the integration scheme and compared with the case of TSV-free integration. The obtained results clearly

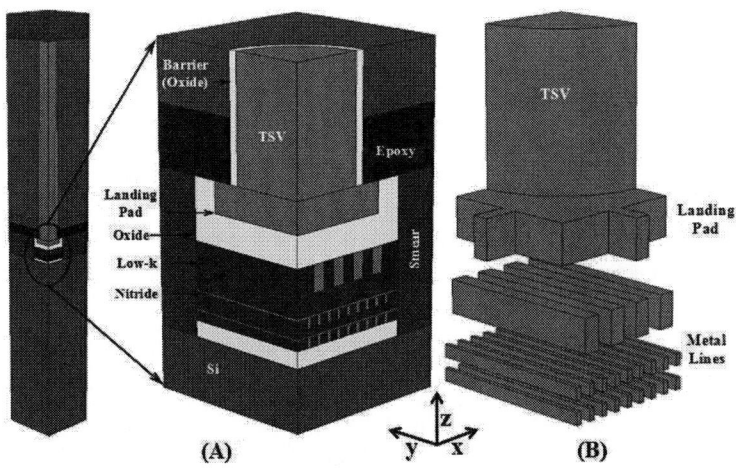

Fig. 5. (A) Global Si/TSV structure and submodeling, (B) Landing pad and metal lines in the submodel (back view) [11].

indicated that the presence of TSVs altered the mechanical stress in the structures and led to the formation of a region with higher stress. The mechanical stress distribution in BEoL interconnects was dependent on the TSV material. The impact the mechanical stress distribution on the reliability was assessed by introducing defects in the studied structures. It was concluded that even a small defect in a 3D IC integrated structure can cause large stresses that can lead to de-bonding and delamination, and that the defect behavior and the mechanical reliability of the 3D IC stack are modulated by the material properties, e. g. Young's modulus, Poisson ratio and Coefficient of Thermal Expansion.

2.2. SM and EM induced failures in TSVs and TSV-based 3D IC BEoL/BRDL interconnects

Not so many results from EM studies in 3D TSV stacked systems have been reported so far. The reason seems to be the general believe that since the current density in TSVs is smaller than in the BEoL interconnect segments, the failure should be happened firstly in these segments. However, this is not always true. As mentioned above, a complex picture of the stress distribution is generated inside the thinned dies mounted in a 3D IC stack. These non-uniform stress distributions generated by thermal ramps cause the migration of atoms/vacancies that can generate void nucleation and growth even before an electrical current is applied. An essential current crowding at corners, where TSVs are attached to the landing pads, and the accompanying increase in Joule heat release, that happens preferably at these locations, can be responsible for a degradation process and a failure development. This kind of consideration was used in the analysis of the EM performance of TSVs in interposers reported in [15-17]. Chen et al. [16] modeled EM in a TSV structure with landing pads, but they did not consider the TSV effect on wires connected to the landing pad. While mentioning the important role of interfaces and grain boundaries (GB) in the vacancy migration and void nucleation and evolution, they nevertheless did not include it into the simulation scheme. Tan et al. reported in [17] that thermo-mechanical stress gradients were the major root-cause of the TSV failure. The ratio of contributions of different driving forces, such as the "electron wind" and hydrostatic stress and temperature gradients in the atomic flux divergence were presented without an explanation of the employed formalism. The role of interfaces and grain boundaries in the degradation evolution was not considered. In [13], Pak et al. reported a detailed EM modeling for TSV including the BEoL interconnect segments connected to the landing pad. They studied the impact of TSV-induced stress on the EM lifetime of neighboring interconnects. While starting with a correct set of linked partial-differential equations describing the multi-physics character of the problem they misinterpreted some important details making their results, and consequently, some predictions are not completely reliable. One of these details is an addition of the TSV-induced stress gradient to the EM-induced stress gradient generated by the electrical current-induced redistribution of atoms. TSV-induced stress gradients were generated by thermal ramps such as cooling down from the anneal or reflow temperature to the test/use conditions. The lattice defect migration caused by these stress gradients results in an elimination of introduced stress non-uniformities [20]. Thus, the diffusion-induced stress relaxation should be taken into account for an accurate determination of the initial (residual) stress, existing in the metal line before an electrical stressing was applied. Another problem is associated with the introduced so-called time metrics which was used to demonstrate the instances in time when the deviations of vacancy concentration from the equilibrium one had reached a certain threshold needed for void nucleation in different locations along the interconnect segment (5% deviation was used by Pak as such threshold). This approach could be a reasonable, DFM-like check of the voiding probability at different locations inside an interconnect line if isn't a well known fact of the abrupt change of the stress distribution inside metal line caused by void nucleation The analysis should be continued with this new distribution of stress depending on the evolving void size and shape. This analysis can demonstrate that in some cases the nucleated void can rather disappear then propagate through the entire length of the interconnect segment.

A short review of the available publications allows us to conclude that the current development provides an initial basis for SM and EM modeling in 3D TSV structures; however, further studies are needed.

3. EXPERIMENT AND SIMULATION – METHODOLOGY FOR RELIABILITY ASSESSMENT IN 3D IC STACKS

3.1. Multi-scale materials data and database approach

The multi-scale simulation methodology for the reliability assessment, which was developed similarly as for the assessment of stress-induced device performance in 3D IC stacks [21] requires multi-scale materials data with high accuracy too, as input for the simulation. This multi-scale approach is of particular importance since for sub-micron and particularly for sub-100nm scales the materials properties are changing. In addition, materials are changing from technology node to technology node in microelectronics industry – low-k and ultra low-k dielectrics and other dielectrics as well as metallic coatings and barriers are some examples.

Since the scales for the simulation and the respective structures are the same, independently of the fact if the modeling and simulation is done to understand and to improve the device performance or the reliability, the analytical techniques for the determination of multi-scale materials parameters are the same as listed in [21].

The generation of a database for multi-scale material parameters of wafer-level and package-level structures is a practical approach to provide a set of data in a particularly designed format to the modeling and simulation community.

3. 2. 4D multi-scale materials characterization methodology

Reliability engineering includes the methods to prevent or to minimize the occurrence of failures during the lifetime of a product as well as the measurement and the modeling of degradation (time-to-failure, failure rate). Accelerated tests provide failure rates and time-to-failure, i. a. data for lifetime estimates under operation conditions (use conditions) of a product, including the stress-dependence (external agents) on lifetime. In case of electromigration in microelectronic products, accelerated tests are performed at elevated current densities and temperatures, the resulting data are average lifetime t_{50} and activation energy for atomic transport E_a.

Reliability engineering has direct influence on:

- Product design (rules): Electrical and mechanical engineering

- Materials selection (criteria): Materials and mechanical engineering

- Manufacturing processes (guidelines): Process engineering.

However, based on statistical data that characterize the reliability of a product and on post-mortem failure analysis (usually deprocessing and cross-sectioning), no information is provided regarding the physical reliability-limiting processes and kinetics of degradation mechanisms.

Reliability physics studies the kinetics of degradation mechanisms (materials and device degradation) that eventually cause failures of a product. In the case of electromigration, the time dependence of interconnect degradation is investigated, using current density and temperature as parameters. However stress is affecting the electromigration too.

In-situ experiments are necessary to study the time-dependent degradation of interconnects in integrated circuits and 3D stacks of integrated circuits, to understand the reliability-limiting processes in microelectronic products and to proof physics-based models. Tomography experiments (3D) provide the whole spatial information, combined with the time-dependence 4D experiments and particularly 4D microscopy are the best possible approaches.

Considering the multi-scale approach of modeling and simulation, and the respective approach for the analytical techniques to study the kinetic processes, X-ray tomography and electron tomography are the two experimental techniques that are supposed to be applied within the 4D microscopy approach [22].

In addition, these X-ray and electron microscopy techniques can be applied to study adhesive and cohesive failures in microelectronic products, including the crack propagation behavior. A miniaturized double-cantilever beam technique was proposed in [22] for in-situ X-ray crack propagation studies in integrated circuits.

3. 3. Physics-based modeling and multi-scale simulation methodology

Similarly to the multi-scale simulation methodology that was developed for the assessment of the effect of stress on device performance in 3D IC stacks [21], the reliability assessment also requires the development of a multi-scale simulation flow. Indeed, in order to be on a position to predict any stress-induced failure such as cracking or delaminating inside interconnects or around TSVs, stress or EM-induced voiding, etc. an accurate simulation of stress generated by all important stress sources has to be performed. As an example, Fig. 6 demonstrates distributions of a lateral stress inside a bottom die of the stack consisting of a substrate and two dies, simulated for two different cases. In one case, C4 bumps were resolved as stress sources (Fig. 6, a), and in another one just underfill material was accounted (Fig. 6, b). Fig. 7 shows the distributions of the lateral stress along the line crossing the silicon bulk just above the interface between BEoL interconnect and silicon bulk: the blue line is for the case when the presence of an uniform underfill layer was assumed, the red line is for the case when bumps were taken into account. It is easy to see that the difference in the stress values in these two cases can reach 100-200 MPa. This stress difference will play an essential role if, for example, the energy release rate should be calculated for a crack extension, or for the interfacial delamination in a TSV structure. This stress variation should also be considered if the stress relaxation has to be calculated in order to determine the initial stress existing inside interconnect segments when an electrical current is applied.

This simulation methodology/flow should result in the development of a design verification tool that will be capable to analyze any 3D die stack design for the determination of the across-die variations of the mechanical stress generated by die stacking using the 3D TSV technology. The tool can be calibrated for any technology employed by semiconductor fabs, and should use experimentally determined, accurate materials data that are specific for each technology node. Subsequently, any design can be verified with the tool calibrated for the same fab/technology/materials set that will be used for chip manufacturing and 3D stacking. No additional calibration will be required.

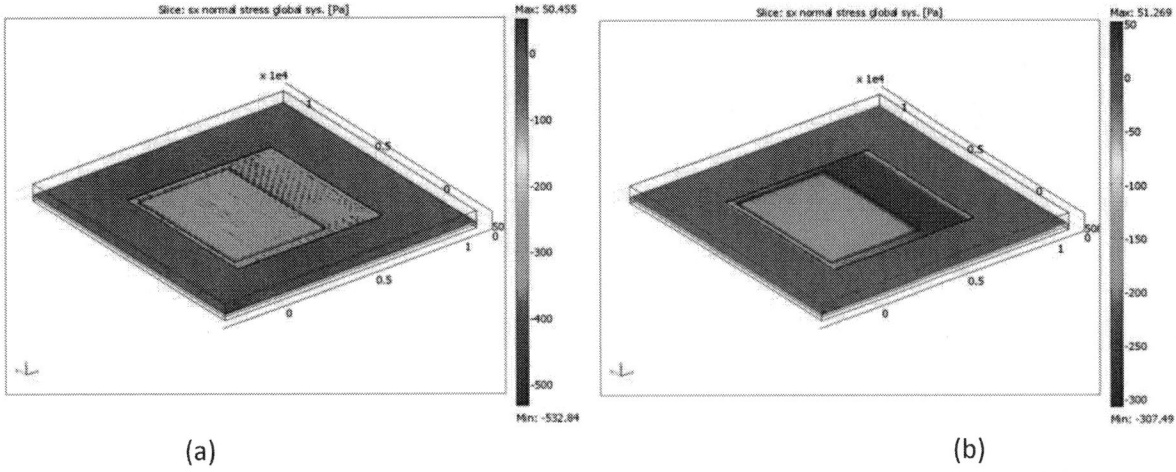

(a)   (b)

Fig. 6. Lateral stress distribution across the bottom die bulk when FC bumps were resolved (a), and were not resolved (b) (stress is in MPa).

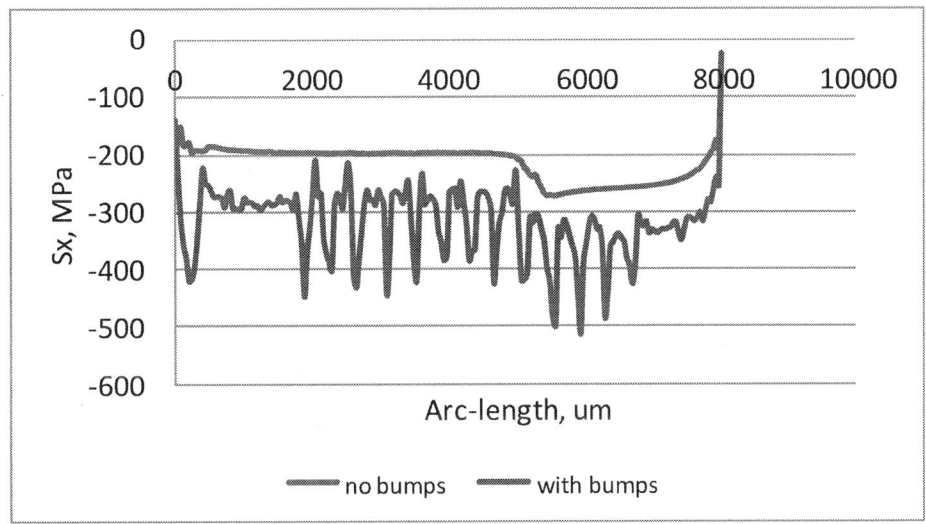

Fig.7. Distributions of the lateral stress along the BEoL lines crossing the silicon beneath the array of FC bumps. Blue line – uniform underfill layers were assumed, red line – bumps were taken into account.

The major target of the proposed simulation flow is the calculation of across-die distributions of stress components for the further assessment of the thermo-mechanical reliability. Known distributions of stress components will allow calculating the driving forces for both cohesive and interfacial crack growth based on fracture mechanics [6, 9, 10], and simulating the stress relaxation inside interconnect metal segments by solving a set of continuity equations, governing vacancy generation/annihilation and migration, coupled with corresponding force balance equations, resulting the strain/stress evolution [23].

The flow is described as the following:

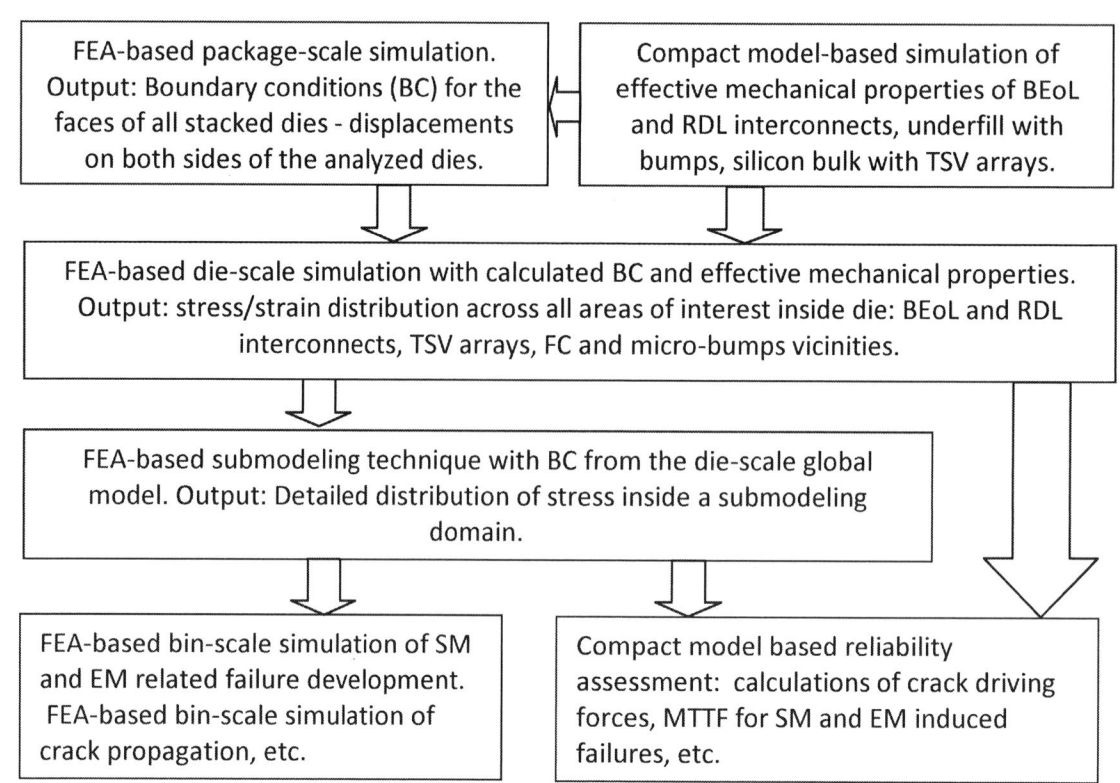

This simulation flow can be described as a sequence of the simulation steps performed with different simulation tools:

A. Calculation of the effective mechanical properties.

Thinned silicon dies with metal TSVs as well as BEoL and BRDL interconnects, and underfill layers with the embedded bumps are approximated as layers with a spatial distribution of elastic properties determined by their layouts. A calculation methodology of the effective Young's modulus and the Poisson ratio as well as the CTE of layers as functions of metal density ρ_M in all metal levels should be developed based on the model for mechanical properties of anisotropic composite materials [24]. This methodology requires a division of all considered composite layers into a number of bins, with the sizes dependent on the required calculation accuracy: the finer partitioning provides more accurate results but for the expense of the run time. For example, for the i-th bin of j-th layer, depending on the routing direction, the Young's modulus should be calculated using one of the following formulas

$$E_{\text{II}}^{i,j} = E_M \rho_M^{i,j} + E_D\left(1 - \rho_M^{i,j}\right) \tag{1}$$

$$E_{\perp}^{i,j} = \frac{E_M E_D}{E_D \rho_M^{i,j} + E_M\left(1 - \rho_M^{i,j}\right)} \tag{2}$$

An example of the across-die distribution of the lateral component of the effective Young's modulus E_{int}^x of BEoL interconnects is presented in Fig. 8. Figure 8,a demonstrates an across-die distribution of the x-component of the effective Young's modulus of BEoL interconnects for the layout (GDS II) shown in Fig. 8,b. An example of a successful implementation of this methodology can be found in [25].

B. Package-scale simulation.

3D IC integration steps such as TSV fill, wafer or die thinning, bump-mounting, further second level packaging, etc. are responsible for the generation of the package-induced stress or CPI. The stresses generated by these stress sources should be resolved at the package-scale simulation step. The output of the FEA-based package-scale simulations should be used for generating a set of boundary conditions (BC) representing the packaging-induced fields of displacements at the surfaces of the already thinned die for the subsequent simulation of the strain distribution everywhere inside the die (Fig. 9). The only difference between the proposed package-scale FEA model and the "standard" one is in the representation of the analyzed die (tier). In the proposed flow, a silicon die represents a solid brick consisting of three layers, which are BEoL interconnect stack, silicon bulk with TSVs, and BRDL, characterized by smeared mechanical properties calculated by averaging the corresponding effective properties over the layer volume.

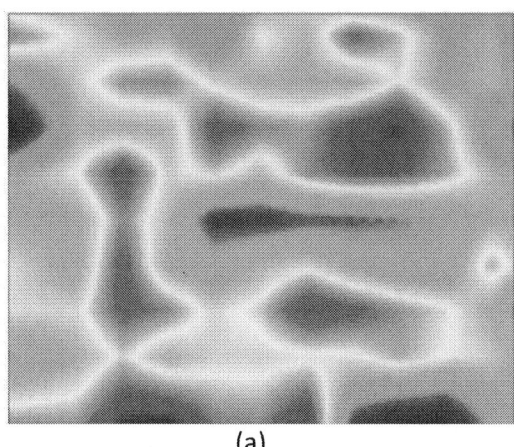
(a)

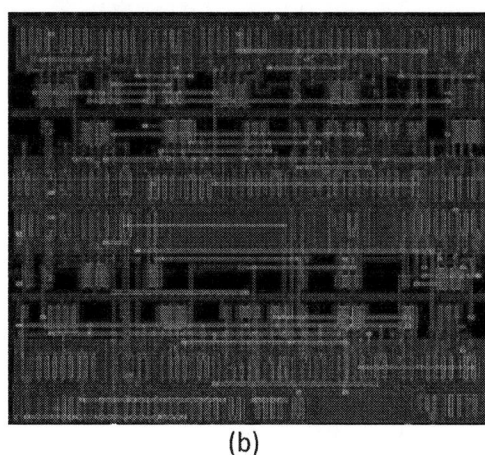
(b)

Fig. 8. Across-die distributions of the x-component of the effective Young's modulus (a) for the multilayer BEoL interconnect layout shown in (b).

C. Die-scale simulation.

The distribution of the CPI/TSV-induced strain inside the die should be calculated using conventional FEA tools. The displacement BC should be taken from the package-scale simulation. BEoL and BRDL interconnects and silicon die/metal TSV layer should be characterized by the spatial distributions of anisotropic effective mechanical properties governed by the case-specific distributions of the metal/dielectric and metal/silicon volume fractions given by the corresponding GDS II layer information (Fig. 8).

D. Detailed distribution of stress simulated with the submodeling FEA-based technique.

If the submodeling is employed, the procedure should be the following: a bin with linear dimensions determined by specifics of the considered applications (SM, EM, cracking, etc.) should be cut off from the die. A proper BC, representing the interaction with the context, surrounding the bin, should be extracted from the previous step of the global modelling (die-scale simulation). The field of displacements on all bin faces can be a good example of these BC. The displacement field solved at the previous step (die-scale simulation) is interpolated onto the submodel as a set of displacement boundary conditions [26]. The validity of the submodels should be confirmed by comparing the displacement field and stress distribution with the global/die-scale models. Therefore, by modeling at different levels, we can retain the information from the global models and transfer this global loading to the smallest scales without loss of information. This technique makes it possible to evaluate the impact of the global packaging process on the local structures (Fig. 10).

35

E. Reliability assessment.

The following reliability assessment can be done by one of two different approaches:

I. FEA-based simulation of degradations and failures.

II. Compact modelling.

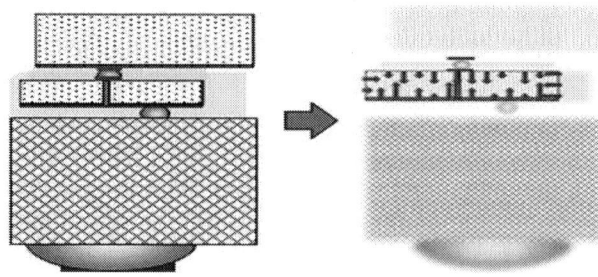

Fig. 9. Displacement BC is an output of the package-scale simulation and an input for the die-scale simulation.

E.1. FEA-based simulation of degradations and failures

Since the size of a submodeling case is small in comparison with the original package-scale size, we can add different physics to the model (if needed) in order to simulate a degradation and failure. For example, if we study the dependency of the TSV-interconnect interaction on the TSV location, we should simulate it by applying the global model first, as discussed above, and subsequently use the calculated displacements as BC for the new, submodel-related simulation domain. If, for example, we simulate SM/EM failures in a BEoL interconnect segment located in the TSV proximity, this displacement BC should be introduced into the corresponding application mode ("Structural Mechanics", in the case shown in Fig. 11) of the simulated multiphysics problem.

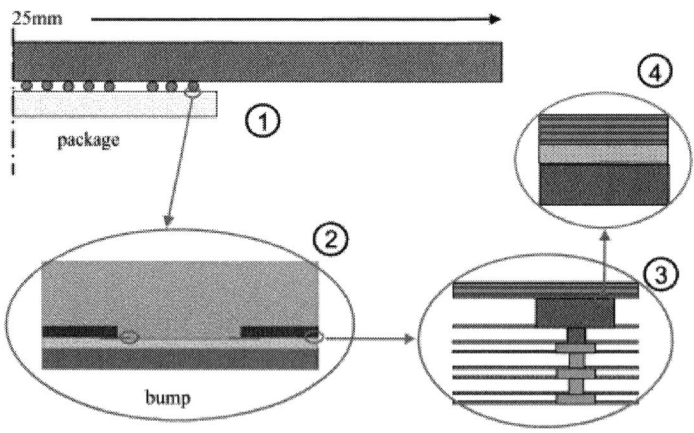

Fig. 10. Multilevel multiscale submodeling technique [26].

Hence, the global FEA model in this case consists of just one application, which is the "Structural Mechanics" with the included thermal load. The submodel consists of at least four different applications such as "Conductive Media (DC)", "Structural Mechanics/Plain Strain", and two additional applications for the solution of partial differential equations (PDE) describing the time-dependent evolution of vacancy and plated atom concentrations [20]. Fig. 11 shows the hydrostatic stress distributions everywhere in the simulated submodel domain for three cases. The first case was simulated for the state of equilibrium which was developed by vacancy generation/annihilation and migration at the anneal temperature of T_{an}=573K (Fig. 11, a). The second stress distribution, shown in Fig. 11, b, was developed due a thermal ramp down from T_{an} to a

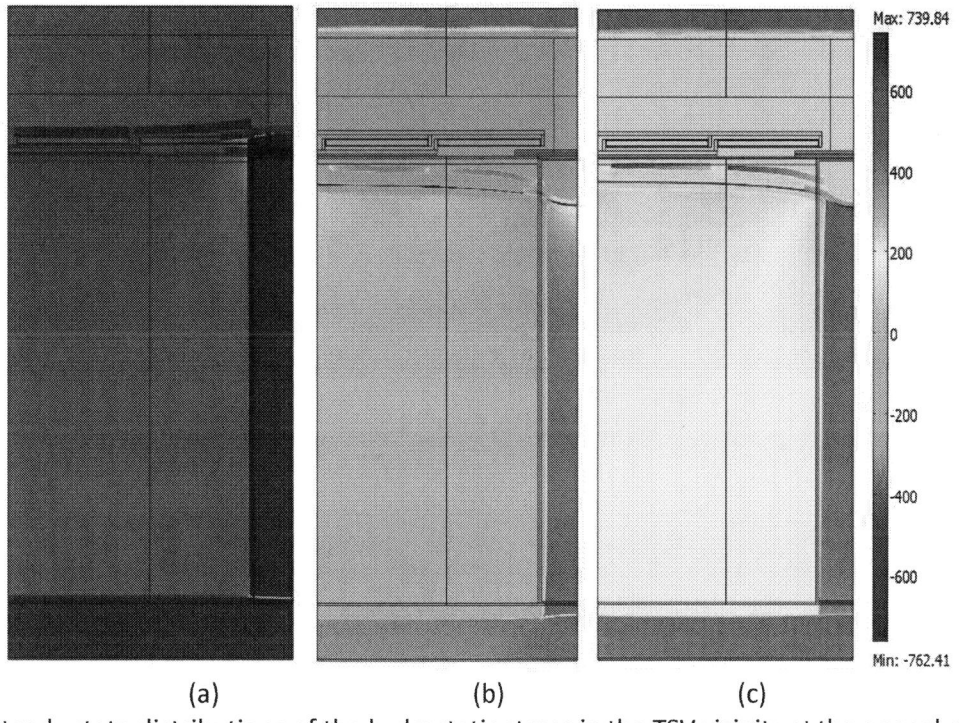

(a)   (b)   (c)

Fig. 11. Steady state distributions of the hydrostatic stress in the TSV vicinity at the anneal stage (a), after the thermal ramp down to a test temperature (b), and under the EM conditions (c), stress in MPa. Deformed shape scale factors were 2000 for (a) and 50 for (b) and (c) cases.

test temperature of T_{test}=373K. Stress-induced vacancy generation/annihilation and migrations were included into simulation. Figure 11, c shows the hydrostatic stress distribution in the same submodel domain when a DC current was passing through TSV (j=8.8x10^{9}A/m2) and through the resolved two interconnect metal lines (j=3.4x10^{10}A/m2) at T=373K (Fig. 12). Figure 13 shows the steady state distributions of the hydrostatic stress along these two lines at T_{an} (Fig. 13,a), after the thermal ramp down to T_{test} (Fig. 13,b), after completion of the diffusion-induced stress relaxation at T_{test} (Fig. 13,c), and, finally, when the EM stressing was applied (Fig. 13,d). It can be seen from these graphs that the

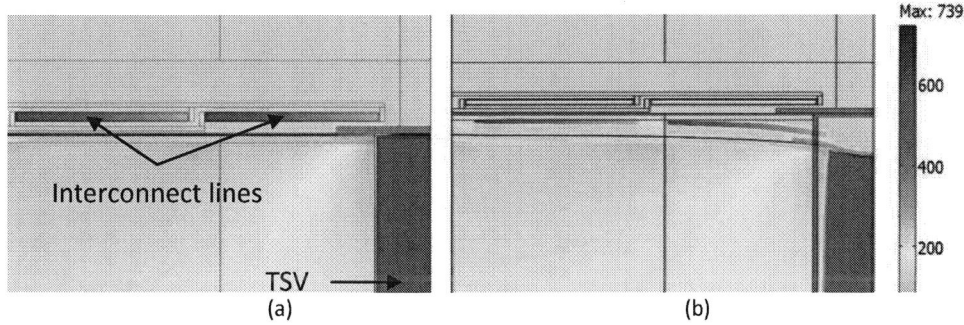

(a)   (b)

Fig. 12. BEoL interconnect metal lines in TSV proximity. Colour map shows the distribution of the hydrostatic stress caused by EM stressing and interaction with TSV (a), and deformations (b); deformed shape scale factors is 50 for (b).

stress gradient induced vacancy generation and vacancy migration result in an uniform hydrostatic stress distribution in each metal line (Fig. 13, a and c). It can be seen that slightly different values of the hydrostatic stresses in these lines, depending on the line positioning relative to the TSV, were generated. It should be noted that the obtained time intervals between the instances in time when the electrical current was loaded and a void was nucleated, which were simulated for several current densities j, are described well by the power functions of j:

$$t_{nuc} \sim \frac{1}{j^n},$$

(3)

It was found that the current density exponents n extracted from the simulation results were slightly different for these two lines. We obtained n=1.88 for the line closer to the TSV and n=1.84 for the further one (Fig. 12). The void nucleation time was determined as the instance in time when the hydrostatic stress has reached for the first time the value of 500 MPa. The current densities employed in this experiment were spread in the interval of [4.9x10^{9}A/m2; 3.4x10^{10}A/m2]. This result supports the fact that n depends on stress. Dashed lines in Fig. 14 shows the comparison between

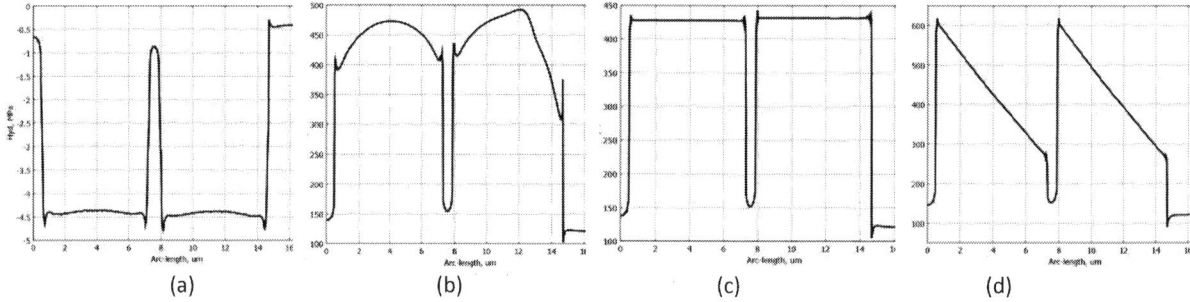

(a)　　　　(b)　　　　(c)　　　　(d)

Fig. 13. Steady state distributions of the hydrostatic stress along two metal lines at T_{an} (a), at T_{test}, after the thermal ramp down from T_{an} to T_{test} (b), after completion of the diffusion-induced stress relaxation at T_{test} (c), and, finally, when the EM stressing was applied (d).

distributions of hydrostatic stress developed in similar line segments located nearby to and far away from the TSV. It is clear that the TSV presence changes the state of stress inside the nearby metal lines toward more tensile, which can be responsible for faster void nucleation.

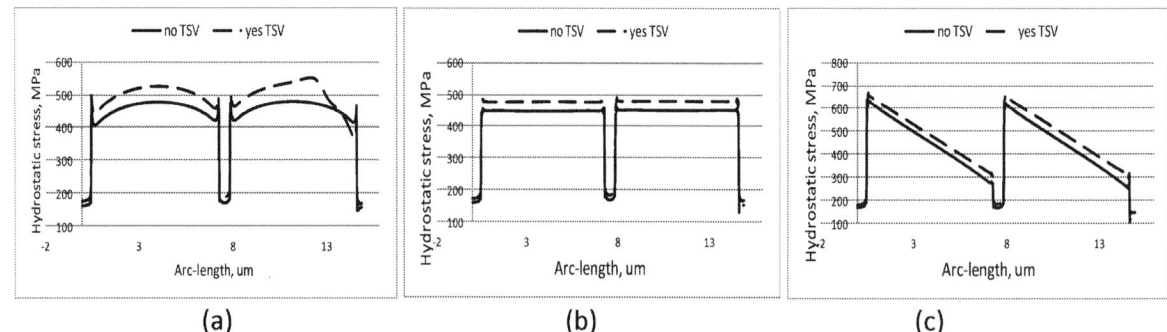

(a)　　　　(b)　　　　(c)

Fig. 14. Steady state distributions of the hydrostatic stress along two metal lines located close to TSV (dashed lines) and far away from TSV, when TSV induced stress is vanished (solid lines) at T_{test}, after the thermal ramp down from T_{ZS} to T_{test} (a), after completion of the diffusion-induced stress relaxation at T_{test} (b), and, finally, when the EM stressing was applied (c).

The demonstrated results were obtained by submodeling a particular TSV. To see how these results depend on the lines location relatively the die edges, the warpages on both faces of the bottom die from the die stack shown in Fig. 6,a has to be calculated. Figure 15,a shows the warpage distributions along a line crossing the die in the direction parallel to the x-axes. Fig. 15,b shows the die thickness variation along that line. It can be seen that the thickness variations can easily be as large as 0.05μm, which produce a strain of 1×10^{-3} (die thickness is 50μm). These displacements when being applied as BC on the top/bottom boundaries of the submodeling domain demonstrate a shift of ~50MPa in all steady state stress distributions simulated for different stages: thermal ramp, stress relaxation, and EM stressing. It also indicates that the critical stress of 500MPa can be reached faster if the tensile stress is generated by packaging.

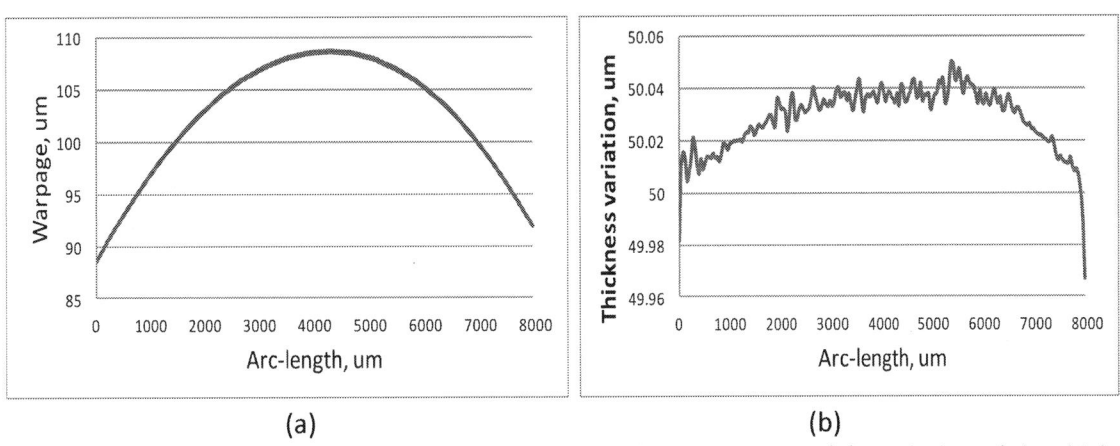

(a)   (b)

Fig. 15. Warpage profile for both faces of the bottom die shown in Fig. 5,a (a). Variation of the thickness of the bottom die, calculated as a difference of warpages for two points with the same lateral coordinates located on the bottom and on the top die surfaces.

The reliability assessment flow described above, when FEA simulations were employed for both scales, i. a. global modeling and submodeling, cannot be employed for the whole die. An extremely large number of submodel simulations that should be done to cover the whole die make this approach unrealistic. A reasonable use-model for such kind of assessment can consist of:
-   FEA-based stress simulation with the global model
-   Detection of the potential hot spots characterized by the excessive stress concentrations
-   Submodeling at these locations.

This approach, while providing a potential capability for the accurate reliability assessment, requires a prior material characterization resulting accurate materials data in order to feed a variety of simulation models. The considered example of SM/EM assessment requires not only mechanical materials data of several materials such as Young's modulus, Poisson ratio, CTE, and, particularly for the assembly and packaging materials, yield stresses, visco-plasticity and creep characteristics, etc. In addition, the resistivity of all involved conducting materials and microstructure-depending characteristics like copper self-diffusivities and diffusivities of copper atoms along interfaces and grain boundaries, depending on the crystallographic orientations of grains, etc. have to be known. For many of these parameters, particularly for assembly and packaging materials, their temperature dependencies have to be determined. Some of these parameters can be derived from accurate materials characterization and some from calibration. A direct calibration procedure should be based on the measured stresses at different locations. A multi-scale stress measurement methodology that mitigates the changes of the original data of the 3D IC system has to be developed/applied in order to prevent a relaxation of stress generated by die stacking.

E.2 Compact model-based simulation of degradations and failures

Another approach to the chip-scale reliability assessment can be developed on the basis of compact modeling. This approach should be based on a DFM-like methodology for the reliability assessment. The chip design community has invented and widely uses the name DFR, which states for "design-for-reliability". Most of the times a DFR-based methodology represents different ways of "designing around" the existing reliability problems. Inevitable penalty for such kind of methodology are an increase in design size, dissipated power and possible timing violations. An alternative approach is the development of a set of verification tools which should allow either to approve the complete design or, by running at the preliminary design steps, to detect all violations leading to reliability risks and eventually failures and to fix them. A number of attempts to develop such verification tools have been reported in the past, and even more attempts are undertaken currently [27-32, 33, 34]. Unfortunately, the majority of the proposed approaches to the reliability assessment has suffered and continues to suffer an absence of reliable models for reliability failures.

For example, in the case of preventing an EM-induced failure of the integrated circuit, a currently employed chip-scale assessment consists of the estimation of a temperature and a current density in each interconnect segment characterized by a noticeable DC current component (mostly are the power/ground nets [28]). If these data are available, a mean-time-to-failure (MTF) is calculated for each segment based on the Black equation [35]

$$MTF = \frac{A}{j^n}\exp\left\{\frac{E_a}{kT}\right\},\tag{4}$$

where j is the current density and E_a is the apparent EM activation energy. Failure times of the large number of the identical interconnect segments are described by the lognormal distribution [36]. The symbol A is a constant, which depends on a number of factors, including grain size, line structure and geometry, test conditions, current density, thermal history, etc. Black determined the value of n equal to 2. However, it is a common understanding today that n is highly dependent on residual stress and current density [37-39] and its value is highly controversial. In the most advanced cases a special kind of segment filtration is performed before the MTF is calculated. This filtration is based on the calculation of the product of electrical current density j and metal line length l for each segment, and the following comparison of the calculated products with the so-called critical product:

$$(j \times l)_{crit} = \frac{\Omega\sigma_{crit}}{eZ\rho},\tag{5}$$

derived from the well-known "Blech limit", [40, 41], which states that any segment characterized by a smaller current-length product than a critical one will be immune to the EM failure. The reason is a balance between the EM-induced atom flow and the counter flow caused by the stress gradient that is buildup by the atom/vacancy redistribution. Here, in (5), Ω is the atomic volume; e is the electron charge, eZ is the effective charge of the migrating atoms, ρ is the electrical resistivity, k is the Boltzmann's constant; T is the absolute temperature; σ_{crit} is the critical stress needed for the failure nucleation (void/hillock). After sorting out all immortal segments, the MTF is calculated for the remaining segments and the minimal MTF represents the MTF for the whole die (weakest-link approach).

In reality, as it was mentioned above, the situation is not as simple since the observed dependency of n and E_a on j and T, the Black equation that was calibrated at stress condition cannot be used for the accurate estimation of MTF at the test or use conditions. Strictly speaking, these inter-dependencies of n and E_a on j and T undermine the validity of the Black equation by itself [42]. The pre-existence of the initial (residual) stress inside the metal segments and, what is most important, the segment-to-segment variation of the residual stress makes the "critical product" filtration not straightforward. We need to evaluate the $(j \times l)$ product for the given interconnect segment against the variable "critical product"

$$(j \times l)^i_{crit} = \frac{\Omega \Delta \sigma^i}{eZ\rho}; \; \sigma^i_{init} + \Delta \sigma^i = \sigma_{crit}, \tag{6}$$

where σ_{init} is the residual stress inside considered interconnect segment. Hence, the "critical product" is not a constant anymore but a variable which depends on segment location, relatively the die edges (global model), and on the local context. Thus, a proper chip-scale EM assessment, in addition to today's requirement of availability of temperature and current density in each interconnect segment characterized by a DC current component, requires a prior knowledge of the residual stress existing inside each analyzed segment. A new, physics-based MTF compact model, which is free of the discussed flaws, has to be developed.

One possible method of calculating the initial stress, as discussed above, is the employment of FEA-based simulation of stress in a global model and in the following submodeling. While being capable of providing the accurate results, this method cannot be considered as a realistic one for the chip-scale EM assessment because of the very large size of the model. Another method is based on a combination of the same FEA-based global model and a compact model-based estimation of the post-relaxation stress inside the interconnect segments. While it is being an approximate, this method looks promising for performing the whole chip EM assessment. The pre-relaxation stress components inside each interconnect segment can be estimated on the basis of the global model. As described above, this model resolves the thermo-mechanical stresses, caused by the thermal ramp, everywhere inside the analyzed chip by introducing the BC extracted from the package-scale thermo-mechanical analysis and employing the coordinate-dependent effective mechanical properties for BEoL and BRDL interconnects, and silicon/TSV bulk. Stress inside a particular interconnect segment, such as a metal line, is calculated by averaging the stress values for all nodes of the global mesh located inside this segment (Fig. 16,a). If the linear sizes of the segment are small in comparison to the mesh size, the stress inside this segment is determined by a linear interpolation of the stresses in the neighboring nodes (Fig. 16,b).

If the intra-segment stress generated by the thermal ramp and the package-induced stress is known, its relaxation caused by the accommodation of the vacancy concentration to this new state of stress has to be calculated. The easiest way to estimate a scale of this accommodation is to consider a single copper grain embedded in a rigid confinement. This grain should be a subject of rapid change in the interior stress caused, for instance, by a rapid thermal impact. The change in the state of stress activates a generation/annihilation of vacancies at the grain boundary and a subsequent diffusion of vacancies into/from the grain interior. Following Herring [20], we model a grain as a spherical object of a radius R with a subsurface region of the thickness δ representing a grain boundary, along which the generation/annihilation of the vacancy-plated atom pairs occurs. Thus, at the zero stress condition the vacancies are distributed uniformly across the grain with a thermodynamically equilibrium concentration N_0. An introduced stress σ_T disrupts the equilibrium of the vacancy concentration. It activates the vacancy migration and the generation/annihilation of vacancy-plated atom pairs. Plated atoms, which, since their low diffusivity in comparison to the vacancy diffusivity, are assumed to be immobile, generate a compression while the tensile stress is generated by vacancies [23]. The solution of the standard force balance equation, with the additional terms related to the dilatational strains, caused by vacancies and plated atoms, provides estimates of the new steady state concentrations of vacancies and plated atoms, and the hydrostatic stress values at the grain boundary (GB) and in the grain interior (GR):

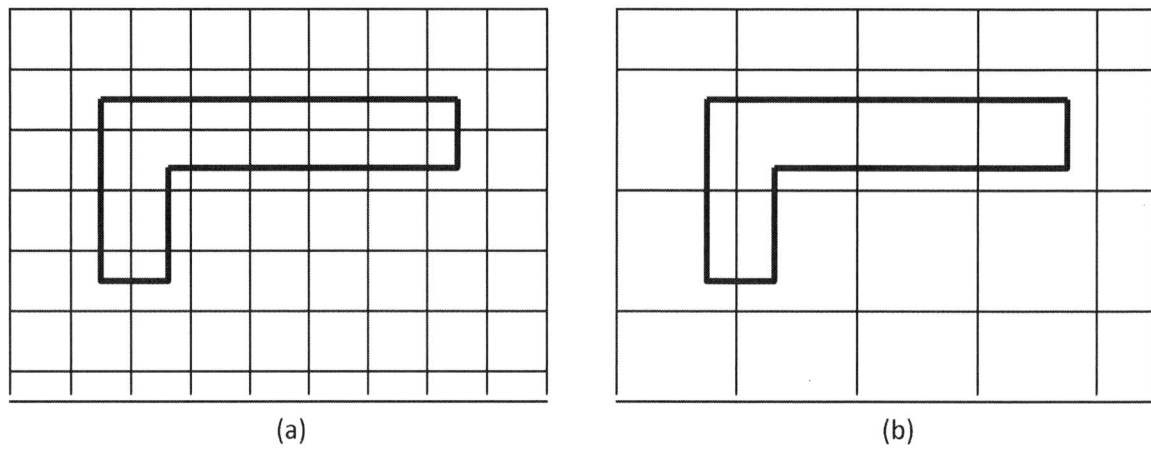

(a)   (b)

Fig. 16. Schematic representation of the interrelation between the global model grid size and the linear sizes of interconnect segment in two cases: the grid size is smaller (a) and larger (b) the minimal linear size of segment.

$$\sigma_{Hyd} = \begin{cases} \sigma_{Hyd}^{GR} \approx \sigma_T + \dfrac{2}{3}\dfrac{E}{(1-\upsilon)}\left(N_0^{ZS}-N\right); & 0 \leq r \leq R \\ \sigma_{Hyd}^{GB} \approx \sigma_T + \dfrac{2}{9}\dfrac{E}{(1-\upsilon)}\left(\dfrac{R}{\delta}\right)\left(N_0^{ZS}-N\right); & R \leq r \leq R+\delta \end{cases}$$

$$N_{GR} = N_{GB} \approx N_0 e^{\frac{f\Omega\sigma_T}{kT}}\left(1+\dfrac{2}{9}\dfrac{\Omega}{kT}\dfrac{E}{1-\upsilon}\left(\dfrac{R}{\delta}\right)\left(N_0^{ZS}-N_0 e^{\frac{f\Omega\sigma_T}{kT}}\right)\right) \approx N_0 e^{\frac{f\Omega\sigma_T}{kT}}$$

(7)

$$M \approx -\dfrac{1}{3}\left(\dfrac{R}{\delta}\right)\left(N_0^{ZS}-N_0 e^{\frac{f\Omega\sigma_T}{kT}}\right)$$

Here, $N_0 = \exp\{-E_A/kT\}$ is the equilibrium vacancy concentration at temperature T, N_{GB} and M are the concentrations of vacancies and plated atoms at GB, while N_{GR} is the concentration of vacancies in the grain interior; σ_{Hyd} is the hydrostatic stress; $f = \Omega_v/\Omega$ is the ratio of the volume occupied by a relaxed vacancy to the atomic volume; E and ν are the copper Young's modulus and Poisson ratio. Obtained results indicate that the hydrostatic stress is more tensile or less compressive, depending on the nature of σ_T, at the GB in comparison with the grain interior. Estimations of the scale of stress relaxation caused by lattice defects equilibration with the temperature and stress provides for GB $\Delta\sigma_{Hyd}^{GB} \approx 20 \div 200 MPa$ and for grain interior of the order of *100 kPa*, which are valid for the ratio (R/δ) of $10^3 \div 10^4$. Excessive tensile stress in the GBs is caused by a larger number of vacancies annihilated with plated atoms due to difference in their equilibrium concentrations at T_{ZS} and at test temperature T in comparison with the number of generated vacancies due to a tension developed by cooling down from T_{ZS} to T. Figure 17 shows the schematics of the equilibrium distributions of hydrostatic stress and the concentration of vacancies along the grain radius. Despite of the idealized nature of the considered case, there is a hope that it demonstrates a true character of the expected relaxation. The validation of the

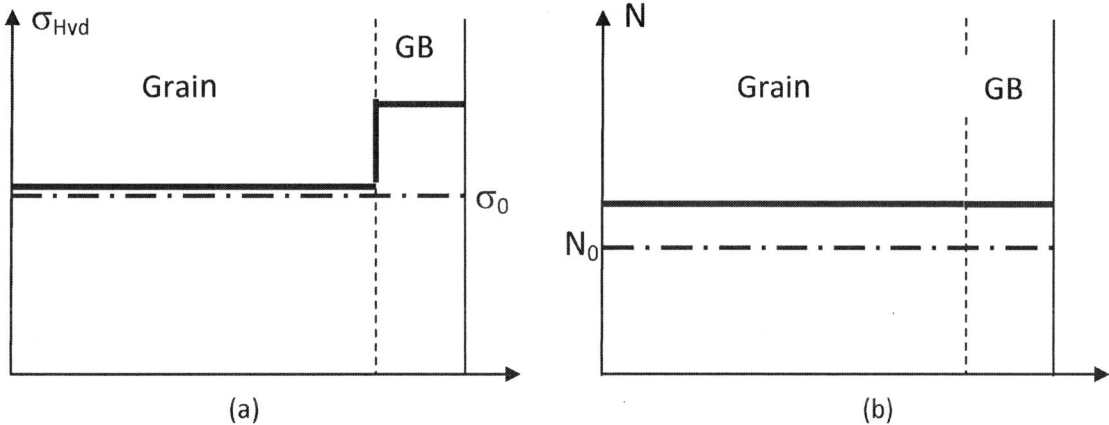

(a)   (b)

Fig. 17. Schematics of the equilibrium distribution of the hydrostatic stress across a grain (a) and the vacancy concentration (b), after accommodation of the stress σ_0.

derived analytical expressions describing the steady state hydrostatic stress and concentrations of vacancies and plated atoms in the considered grain/grain model, shown in Fig. 18, was done by comparison with the results of the FEA simulation. A multi-physics SM simulation setup [23, 43] was used. The simulation results demonstrate evolution of the radial distributions of vacancies, plated atoms and stress. Steady state uniform distributions are achieved at times of order of 10^5s. The kinetics of this evolution is controlled by diffusion of vacancies from the grain interior to GB where its annihilation with the plated atoms takes place. This process reduces the high initial concentration of vacancies to smaller concentration corresponding to the test temperature and increases the stress developed in the GB by cooling and defect induced dilatation.

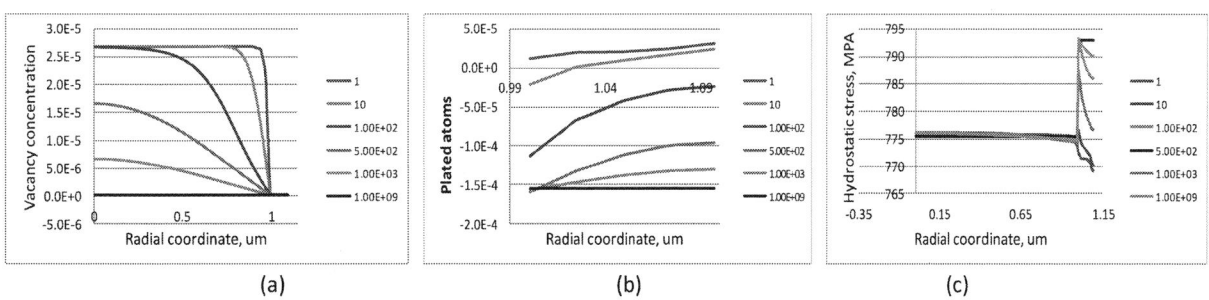

(a)   (b)   (c)

Fig. 18. Kinetics of the radial distribution of the vacancy concentration (a), plated atom concentration (b), and the hydrostatic stress (c). Time is in seconds.

Hence, now we can assume that the initial state of stress, generated by the thermal ramp down from the anneal temperature to test temperature and by a diffusion-induced stress relaxation, is known. Thus, we have everything needed for the modeling the next step, which is a stress redistribution caused by an applied electrical load.

It is a common understanding that the EM phenomenon consists of a redistribution of lattice atoms and lattice defects caused by a directed electrical current. This redistribution is accompanied by a noticeable change in the state of stress: a buildup of the hydrostatic stress gradient characterized by a tensile stress generated near cathode and a compressive stress generated near anode. As it was shown before, see for example [44, 45, and 23], the steady state developed during EM stressing is characterized by a linear distribution of the hydrostatic stress and an exponential distribution of the vacancy concentration along the interconnect metal line (segment):

$$\hat{\sigma}(x) = \sigma_T^R - \frac{eZ\rho jx}{2\Omega}$$

$$\hat{N}(x) = N_T^R e^{-\frac{eZ\rho jx}{2kT}}$$

$$\hat{M}(x) = M_T^R + \frac{eZ\rho jx}{2\Omega B}$$

(8)

Here, $\hat{\sigma}, \hat{N}, \hat{M}$ are the equilibrium distributions of hydrostatic stress as well as vacancy and plated atom concentrations, B is the effective bulk elastic modulus. σ_T^R, N_T^R, M_T^R are the same characteristics, existing in the considered interconnect metal segment before an electrical current was applied. These characteristics were described in (7) as σ_{Hyd}^{GB}, N, and M. All considered characteristics are taken for GB/interface regions since the performed analysis predicts the biggest hydrostatic stress to be developed at the metal line interfaces at the cathode/anode ends, Fig. 19, b.

The results of the FEA simulation of the evolution of hydrostatic stress and concentrations of vacancies and plated atoms demonstrate that the long-term kinetics of EM induced stress relaxation, similarly to considered case of stress-induced relaxation, depends on the vacancy exchange between grain interior and GB. These results were obtained from the simulation of the stress/concentrations evolution kinetics caused by DC stressing in the model system represented by two rectangular grains of length *l* and diameter *2R*, surrounded by thin GB layer of the thickness *δ*, and embedded into the rigid confinement (Fig. 19).

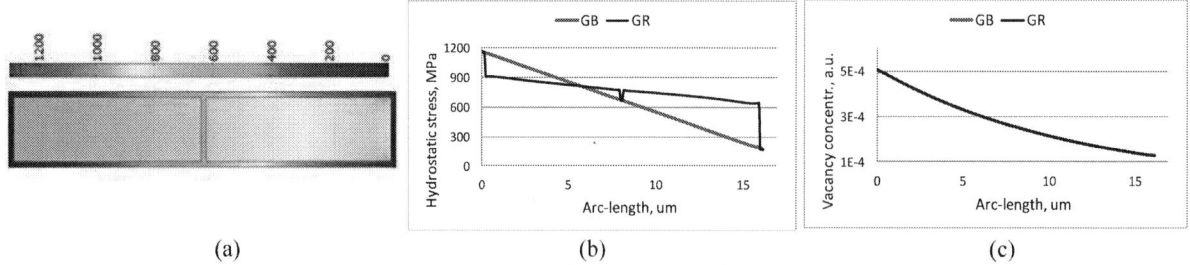

(a)   (b)   (c)

Fig. 19. Hydrostatic stress distribution in two-grain model (a); steady-state distributions of the hydrostatic stress (b), and vacancy concentration (c) in the GB and grain interior.

A simple analysis of the evolution of the GB concentrations of vacancies and plated atoms at the instances in time when the system is in the vicinity of equilibrium has generated an analytical formulation of the void nucleation time. This critical time was determined as a moment in time when the hydrostatic stress reached for the first time the critical value, which was 500 MPa in the considered example. The void nucleation time is given by

$$t_{nuc} \approx \tau \frac{kT_{test}}{\Omega B} \frac{1}{\hat{N}_C} \ln\left\{\frac{B\left(\hat{M}_C - M_T^R\right)}{\sigma_T^R - B \cdot \hat{M}_C - \sigma_{CR}}\right\} \tag{9}$$

Here, $\hat{N}_C$, $\hat{M}_C$ are the steady state vacancy and plated atom concentrations at the cathode edge of the metal line (the electrons inlet), where the hydrostatic tension is maximal; $\sigma_T = f\left(\vec{r}, \Delta\alpha(T_{ZS} - T)\right)$ is the global stress representing the combination of the thermal and CPI-induced stresses calculated in the global model, T_{ZS} is the "zero stress" temperature, T is the test temperature, and $\Delta\alpha$ is the difference in CTEs of all involved materials; σ_{CR} is the critical hydrostatic stress needed for void nucleation; τ is the time constant effecting the vacancy/plated atom generation/annihilation kinetics [46, 44, 43], which is proportional to the time of vacancy diffusion along the line length L:

$$\tau = \frac{L^2}{D_{V0}e^{-\frac{E_D}{kT}}}, \tag{10}$$

where E_D is the activation energy for the vacancy diffusion. Substituting all mentioned above parameters in the equation (10) by their expressions given by (7), (8) and (10) provides

$$t_{nuc} \approx \frac{L^2}{D_{V0}} e^{\frac{E_V + E_D}{kT}} \frac{kT}{f\Omega B} \exp\left\{-\frac{f\Omega}{kT}\left(\sigma_T + \frac{B}{3}\left(\frac{R}{\delta}\right)e^{-\frac{E_V}{kT_{ZS}}}\right) - \frac{eZ\rho l}{4kT}j\right\} \times$$
$$\times \ln\left\{\frac{\dfrac{eZ\rho l}{4\Omega}j}{\sigma_T + \dfrac{B}{3}\left(\dfrac{R}{\delta}\right)e^{-\frac{E_V}{kT_{ZS}}} + \dfrac{eZ\rho l}{4\Omega}j - \sigma_{CR}}\right\} \tag{11}$$

Here, as before, j is the DC current density, and T is the test temperature, l is the metal line length, and R is the grain size. It should be mentioned that the void nucleation time extracted from the Korhonen's model [44] is described by almost same analytical formula as (11). Indeed, as it follows from the analytical formulation proposed by Korhonen [44] and further developed by other researchers, see for example [45, 47, 48], the atomic flux divergence results in the volumetric strain, which can be calculated based on the one dimensional diffusion-like equation

$$\frac{\partial\sigma}{\partial t} = \frac{\partial}{\partial x}\left[\kappa\left(\frac{\partial\sigma}{\partial x} + \frac{eZ\rho j}{\Omega}\right)\right] \tag{12}$$

Here, $\kappa = D_a B\Omega/k_B T$, where D_a is the atomic diffusivity, and B is the bulk modulus [44]. Solution of this initial-boundary value problem for the metal wire of the length L with the impermeable boundaries at the ends (diffusion blocked ends) is the infinite series [44, 49]:

$$\sigma = -\frac{eZ\rho j}{\Omega}\left(x + 4L\sum_{n=0}^{\infty}\frac{\cos\left(m_n\left(1/2 + x/L\right)\right)}{m_n^2 \exp\left(m_n^2\kappa t/L^2\right)}\right) \qquad (13)$$

Here, $m_n = (2n+1)\pi$ and $x \in [-L/2; L/2]$. Stress at the wire cathode end ($x = -L/2$) at long time is described by the slowest decaying term of the series (13):

$$\sigma = \frac{eZ\rho jL}{\Omega}\left(\frac{1}{2} - 4\pi^2 \exp\left\{-\frac{\kappa t}{\pi^2 L^2}\right\}\right) \qquad (14)$$

Remembering that the EM induced stress needed for generating a failure is $\sigma_{crit} - \sigma_{in}$, and that it develops at instant in time of $t = t_{nuc}$, we can get from the equation (14):

$$t_{nuc} = \frac{\pi^2 L^2}{D_a}\frac{k_B T}{\Omega B}\ln\left\{\frac{4\pi^2\dfrac{eZ\rho jL}{2\Omega}}{\sigma_{in} + \dfrac{eZ\rho jL}{2\Omega} - \sigma_{CR}}\right\} \qquad (15)$$

As it was discussed above, in the considered case of the vacancy mechanism of atomic diffusion, the effective coefficient of the atom diffusion depends on the vacancy concentration. The latter generates a simple relation between diffusivities of atoms and vacancies

$$D_a \approx ND = D_0 \exp\left\{-\frac{E_V + E_{VD}}{k_B T}\right\}\exp\left\{\frac{\Omega\sigma_{in} + \dfrac{eZ\rho jL}{2}}{k_B T}\right\} \qquad (16)$$

Supplementing D_a in the writing (15) with the expression (16) provides us with the same writing for t_{nuc} as (11), which was derived on the basis of our formulation, dealing with the vacancy-plated atom pairs..

We want to stress that the void nucleation time described by equation (11) is determined by three major stresses. It is the coordinate (intra-layout) dependent combination of the thermal and CPI-induced stresses calculated with the FEA global package model

$$\sigma_T = f\left(\vec{r}, \Delta\alpha(T_{ZS} - T)\right), \qquad (17)$$

This stress increases with increase the gap between the "zero stress" and test temperatures. It is also the diffusion-induced relaxation of the initial σ_T stress

$$\Delta\sigma = -\frac{B}{9}\left(\frac{L}{\delta}\right)\exp\left\{-\frac{E_V}{kT_{ZS}}\right\} \qquad (18)$$

The effective bulk modulus B, which has replaced the multiplier $\dfrac{2}{9}\dfrac{E}{1-\nu}$ used in (7), was introduced with intention to compensate the difference between the ideally spherical shape of the grain, used for the development of (7), and the real grain shape. Thus,

$$\sigma_{Res} = \sigma_T + \frac{B}{9}\left(\frac{L}{\delta}\right)\exp\left\{-\frac{E_V}{kT_{ZS}}\right\} \tag{19}$$

is the residual stress which existed in the considered segment at the moment when the DC current was loaded. And finally,

$$\sigma_{EM} = \frac{eZ\rho l}{4\Omega}j \tag{20}$$

is the additional EM-induced steady-state stress generated at the cathode end of the line loaded with the current density j.

A simple analysis of equation (11) reveals that an interrelation between three major parameters, namely: j, $\sigma_T = f\left(\vec{r}, \Delta\alpha(T_{ZS}-T)\right)$ and σ_{CR}, governs the void nucleation kinetics. For $\sigma_{CR} > \sigma_{Res} + \sigma_{EM}$, which is happen if the condition

$$j \times l < \frac{4\Omega}{eZ\rho}\left(\sigma_{CR} - \sigma_{Res}\right) \tag{21}$$

is valid, equation (11) yields a divergence of the t_{nuc}, which is similar to the Blech condition of immortality [40, 41]. The only difference is the presence of the residual stress: the larger the initial tensile stress preexisting in the metal, the smaller product of the current density and metal line length is required to insure the immortality of the metal line. Another situation occurs for $\sigma_{Res} > \sigma_{CR}$. In this case, equation (11) yields negative void nucleation time. It is obvious that $\sigma_{Res} > \sigma_{CR}$ is the condition for stress voiding. Indeed, if the residual stress exceeds the critical stress needed for void nucleation, then the void will be nucleated before any electrical stressing is applied.

The universal character of the derived formula for the void nucleation time can be further supported by the direct calculation of t_{nuc} for several current densities and test temperatures. The typical values of the parameters that were used in calculations are the following: $L = 10^{-6}$ m; $\delta = 10^{-9}$ m; $E_V = 1.84 \times 10^{-19}$J; $D_{V0} = 7.56\ 10^{-5}$; $E_D = 1.6 \times 10^{-19}$J; T_{test} = 323K - 673K; T_{ZS} = 523 - 823K for σ_{crit} =500MPa; Z =10; ρ =3x10^{-8} Ωm; l =10^{-4}m; $j = 10^9 - 2 \times 10^{10}$ A/m^2; $B = 10^{11}$N/m^2; $\Delta\alpha$ =1.7x10^{-5}1/K. In order to complete the proposed set of calculations, the global stress $\sigma_T = f\left(\vec{r}, \Delta\alpha(T_{ZS}-T)\right)$ was approximated by $B\Delta\alpha(T_{ZS}-T)$, which represents a stress caused by the difference in the copper and silicon oxide CTEs. Figure 20 demonstrates the variation in t_{nuc} calculated for different j and T_{test}. The right upper corner of the $j \times T_{test}$ table demonstrates the immortality region (IMMORT), characterized by the divergence in t_{nuc}. The first left column, corresponding to the test temperature of 323K, shows the negative values for t_{nuc}, i. e., voids had been nucleated during cooling down to T_{test} before electrical stress was applied. The

47

representation of the calculated t_{nuc} in the form of the Black equation and extraction of the traditional current density exponent – n and apparent activation energy E_a (Fig. 21), demonstrate the expected dependencies of n and E_a on T_{test} and j (Fig. 22). The comparison of the predicted dependency $n(T_{test})$ with the measurement results [50, 51], which is shown in Fig. 22, a, demonstrates a reasonable fit. It can be seen that a simultaneous increase of T_{ZS} and σ_{CR} can provide smaller n values. The increase in the activation energy E_a, caused by the increase of the current density (Fig. 22, b), corresponds to the experimental observations [51].

j\Ttest	323K	373K	423K	473K	523K	573K	623K	673K
1.0E+09	T-void	IMORT	IMORT	IMORT	IMORT	IMORT	IMORT	IMORT
2.0E+09	T-void	5.25E+17	IMORT	IMORT	IMORT	IMORT	IMORT	IMORT
3.0E+09	T-void	2.41E+17	4.34E+14	IMORT	IMORT	IMORT	IMORT	IMORT
4.0E+09	T-void	1.35E+17	2.12E+14	1.42E+12	IMORT	IMORT	IMORT	IMORT
5.0E+09	T-void	8.28E+16	1.25E+14	7.09E+11	1.39E+10	IMORT	IMORT	IMORT
6.0E+09	T-void	5.35E+16	8.02E+13	4.33E+11	6.84E+09	3.21E+08	IMORT	IMORT
7.0E+09	T-void	3.58E+16	5.39E+13	2.86E+11	4.27E+09	1.48E+08	1.64E+07	IMORT
8.0E+09	T-void	2.46E+16	3.74E+13	1.98E+11	2.89E+09	9.34E+07	5.97E+06	IMORT
9.0E+09	T-void	1.72E+16	2.65E+13	1.41E+11	2.05E+09	6.44E+07	3.78E+06	3.94E+05
1.0E+10	T-void	1.21E+16	1.91E+13	1.03E+11	1.49E+09	4.65E+07	2.64E+06	2.48E+05
1.1E+10	T-void	8.70E+15	1.40E+13	7.63E+10	1.11E+09	3.45E+07	1.93E+06	1.75E+05
1.2E+10	T-void	6.29E+15	1.03E+13	5.72E+10	8.40E+08	2.61E+07	1.46E+06	1.30E+05

Fig. 20. Calculated t_{nuc} as a function of j and T_{test} under the assumption of σ_{CR} =500MPa.

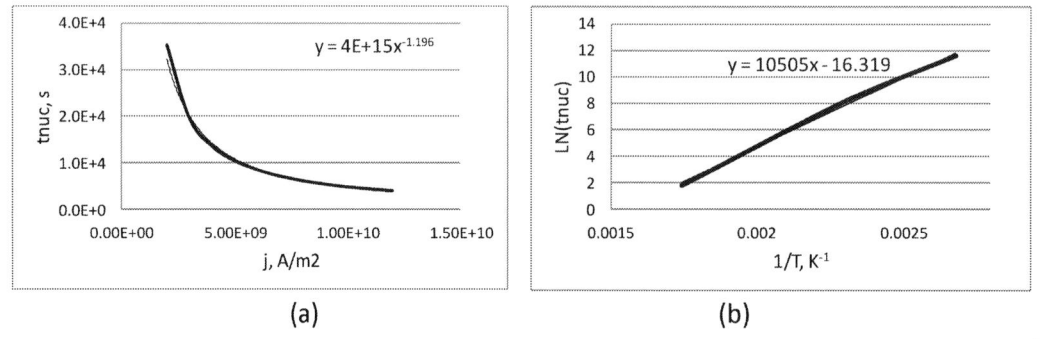

(a)   (b)

Fig. 21. Extraction of the current density exponent n (a) and the apparent activation energy E_a (b) based on calculated t_{nuc}.

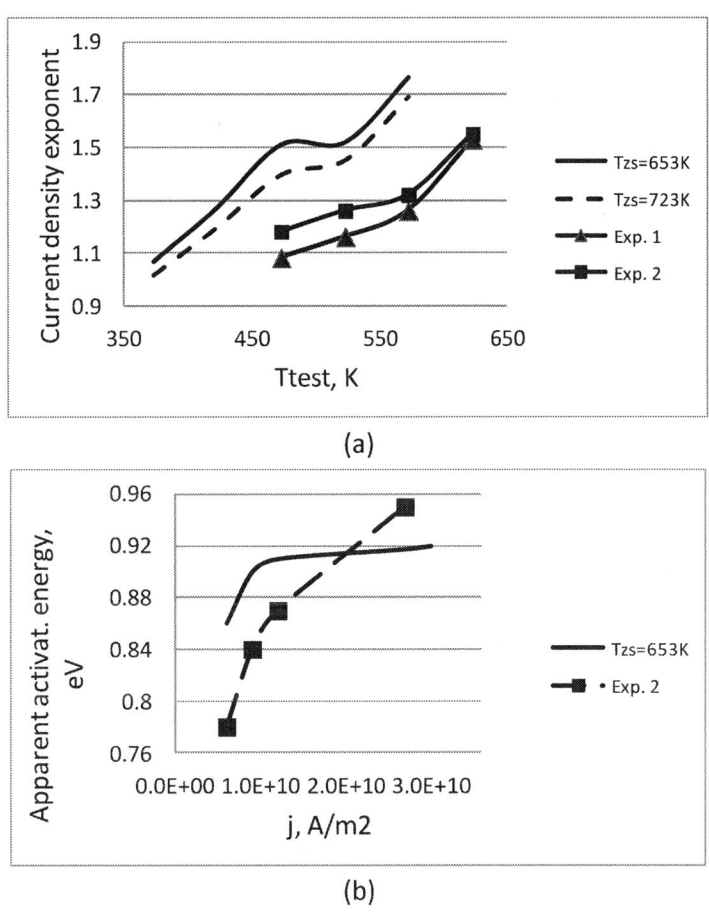

(a)

(b)

Fig. 22. Extracted dependencies of n on T_{test} (a), and E_a on j (b) for σ_{crit}=600MPa vs. experimental data: Exp. 1 data from [50], and Exp. 2 data from [51].

Thus, the scope of the obtained results allows us to conclude that the derived compact model provides a capability for predicting the void nucleation time in the interconnect metal segments taking into account the residual stress generated by a combination of thermal treatment and diffusion-induced relaxation. It combines the predictabilities of the Black equation and the Blech limit, while avoiding possible errors caused by the variability of the current density exponent n and the apparent activation E_a energy, which are treated as constants in the tradition EM assessment.

The final prediction of the MTF requires a detailed knowledge of the mechanism of void nucleation. The employment of the critical hydrostatic stress σ_{CR}, which was used above for the estimation of the void nucleation time, still needs to be proven. Alternatively, the critical value of the normal component of the total stress along the copper interconnect /dielectric capping interface was proposed as a criterion for void nucleation [42]. In this case, interface weakness and delaminating are considered as an initial root causes for void nucleation. Contrarily to this proposal, our previous simulations [43] demonstrated that voids tend to nucleate at interface locations characterized by high shear stress values, Fig. 23. In this case a grain interfacial crack can be easily nucleated.

4.   SUMMARY AND OUTLOOK

In this White Paper, the die-scale EM assessment was demonstrated for a typical example. Despite of the specific character of this example, the proposed methodology for multi-scale simulation of the stress distribution everywhere inside a die can be employed for a wide range of reliability problems where mechanical stress is playing a dominant role. Such relevant examples are calculating the driving forces for both cohesive and interfacial crack growth based on fracture mechanics, prediction of the kinetics of stress-dependent solid state diffusion-controlled reactions, etc. Known distributions of stress components will help to address these problems by determining of hot-spots in the layout, where a degradation evolution is most probable. A subsequent sub-model simulation could provide an understanding of the physical root-cause of the problem.

Another situation might arise when the whole chip assessment is required. This will be happen particularly if appropriate design rules cannot be extracted from the test-chip data due to strong interactions between stress sources. In this case, the submodeling approach cannot be used and a prospective verification tool should examine the whole chip. A proper combination of the FEA-based global analysis and a compact model-based simulation of the particular reliability issue can be a viable option. Thus, a set of compact reliability models, similarly to the existing ones for MOSFET front-end-of-line such as the models for the negative bias temperature instability, time-dependent gate oxide breakdown, and hot carrier injection should be available. Some progress in the development of compact reliability models for the BEoL interconnect stack and particularly for the 3D IC interconnect systems can be mentioned. Examples are the development of a semi-analytical model for the stress distribution near the silicon/metalninterface generated by a TSV [10], and compact models for the early electromigration lifetime estimation [51-53]. All these models have to be validated with experimental data before they can be adopted by Electronic Design Automation tool developers. A probabilistic character of the compact models should be verified against the real large-scale failure statistics.

Another important moment is the availability of the materials properties needed for simulations on all involved scales and experimental data like measured stress components for model calibration. Traditional and novel characterization techniques should be applied or developed for such kinds of measurements/characterizations [21, 54].

Only taken together all discussed the modeling and characterization flows will allow us to design reliable and fast performing 3D ICs.

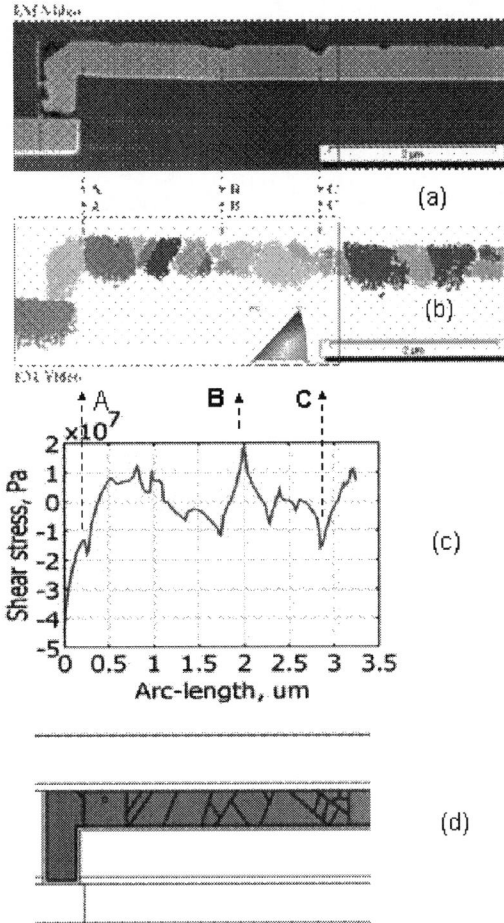

Fig. 23. Post-mortem SEM/EBSD analysis of the sample and simulation results: (a) SEM image of the cross-section of a Cu interconnect segment together with (b) an inverse pole figure (IPF) map (EBSD measurement) of the crystallographic orientations of the individual grains relative to the wafer surface, (c) simulated distribution of the shear stress at the top copper/dielectrics interface for a copper via-line line with the grain structure shown in (d) [43].

References:

1. R. Radojcic, M. Nowak, M. Nakamoto,"TechTuning: Stress management for 3D Through-Si-Via stacking technologies", *AIP Conf. Proc.*, Vol. 1378, pp. 5-20, 2011.

2. V. Sukharev, E. Zschech, "Multi-scale environment for simulation and materials characterization in stress management for 3D ICs TSV-based technologies – Effect of stress on the device characteristics", *AIP Conf. Proc.*, Vol. 1378, pp. 21-49, 2011.

3. M. Murugesan, H. Kino, H. Nohira, et al., „Wafer thinning, bonding, and interconnect induced local strain/stress in 3D-LSIs with fine-pitch high-density microbumps and TSVs", IEDM Tech. Dig. 10-30 (2010).

4. B. Vandevelde, K.J. Rebibis, A. La Manna, et al., Thermo-mechanical impact of the underfill-microbump interaction in 3D stacked integrated circuits, *Electronics Packaging Technology Conference (EPTC)*, 2011 IEEE 13th, 2011, pp. 34-38.

5. I. De Wolg, "Raman spectroscopy analysis of mechanical stress near Cu TSVs", *AIP Conf. Proc.*, Vol. 1378, pp. 138-149, 2011.

6. C. S. Selvanayagam, J. H. Lau, X. Zhang, S. K.W. Seah, K. Vaidyanathan, and T. C. Chai, "Nonlinear thermal stress/strain analyses of copper filled TSV (through silicon via) and their flip-chip microbumps," in *Proc. ECTC*, 2008, pp. 1073–1081.

7. N. Ranganathan, K. Prasad, N. Balasubramanian, and K. L. Pey, "A study of thermo-mechanical stress and its impact on through-silicon vias," J. Micromech. Microeng., vol. 18, no. 7, p. 075 018, Jun. 2008.

8. X. Liu, Q. Chen, P. Dixit, R. Chatterjee, R. Tummala, and S. Sitaraman, "Failure mechanisms and optimum design for electroplated copper through-silicon vias (TSV)," in *Proc. ECTC*, 2009, pp. 624–629.

9. S. Ryu, K. Lu, J. Im, R. Huang, and P. S. Ho, "Stress-induced delamination of TSV structures" in *AIP Conf. Proc.,* Vol. 1378, pp. 153-167, 2011.

10. S. Ryu, K. Lu, X., et Al., , "Impact of Near-Surface Thermal Stresses on Interfacial Reliability of Through-Silicon Vias for 3-D Interconnects", IEEE TDMR, VOL. 11, NO. 1, (2011) pp. 35-43

11. A. P. Karmarker, X. Xu, and V.Moroz, "Performance and reliability analysis of 3D-integration structures employing through silicon via (TSV)," in *Proc. IEEE 47th Annu. Int. Reliab. Phys. Symp.,* Montreal, QC, Canada, 2009, pp. 682–687.

12. X. Xu, A. Karmarkar, "3D TCAD modeling for stress management in Trough Silicon Vias (TSV) stacks", *AIP Conf. Proc.,* Vol. 1378, pp. 53-66, 2011.

13. J. Pak, M. Pathak, S. K. Lim, et al, "Modeling of Electromigration in TSV based 3D IC", IEEE ICTC, 2011, pp. 1420-1427.

14. H. Kadota, R. Kanno, M. Ito, and J. Onuki, "Texture and grain size investigation in the copper plated through-silicon via for three dimensional chip stacking using electron backscattering diffraction", Electrochem. Solid-State Lett., 14 (5) D48-D51, 20011.

15. T. Frank, C. Chappaz, P. Leduc, et al, "Resistance increase due to electromigration induced depletion under TSV", *IEEE Int. Reliab. Phys. Symp. Proc.,* pp. 347-352, 3F.4.6, (2011)

16. Z. Chen, *et al.,* "Modeling of Electromigration of the Through Silicon Via Interconnects", *IEEE International Conference on Electronic Packaging Technology & High Density Packaging (ICEPT-HDP),* 2010, pp. 1221-1225.

17. Y. C. Tan, *et al.,* "Electromigration performance of Through Silicon Via (TSV) - A modeling approach", Microelectronics Reliability, 2010, pp. 1336-1340.

18. V. Sukharev, A. Kteyan, and E. Zschech, "Physics-based models for EM and SM simulation in 3D IC structures", IEEE TDMR, 12 (2012) 272-284.

19. Takana N., Sato T., Yamaji Y., Morifuji T., Umemoto M., Takahashi K., "Mechanical Effects of Copper Through-Vias in a 3D Die-Stacked Module", *Electron. Comp. and Tech. Conf.* 2002, pp.473-479.

20. C. Herring, "Diffusional viscosity of polycrystalline solid", J. Appl. Phys., vol. 21, pp. 437-445, 1950.

21. Stress management for 3D ICs using through silicon vias: international workshop on stress management for 3D ICs using through silicon vias, Editors: E. Zschech, R. Radojcic, V. Sukharev, L. Smith, *AIP Conf. Proc.,* Vol. 1378, pp. 5-20, 2011.

22. E. Zschech, S. Niese, K. B. Yeap, J. Gelb, R. H. Dauskardt, "Mechanical Properties of Cu/ULK Interconnect Stacks Studied with Double Cantilever Beam and Nanoindentation Techniques", *MRS Spring Meeting*, San Francisco, 2012.

23. V. Sukharev, E. Zschech, and W. D. Nix, "A model for electromigration-induced degradation mechanisms in dual-inlaid copper interconnects: Effect of microstructure," J. Appl. Phys., vol. 102, pp. 053505-1-14, Sept. 2007.

24. Jones, RM (1975), *Mechanics of Composite Materials*, Hemisphere Publishing Corporation, New York.

25. V. Sukharev, A. Kteyan, J-H Choy, et. Al, "Multi-scale Simulation Methodology for Stress Assessment in 3D IC: Effect of Die Stacking on Device Performance", J. Electron Test (2012) 28:63–72.

26. Mercado L, Kuo SM, Goldberg C, Frear D (2003), Impact of flip-chip packaging on copper/low-k structures. IEEE Transactions on Advanced Packaging 26: 433-440

27. J. Kitchin, "Statistical electromigration budgeting for reliable design and verification in a 300-MHz microprocessor". In *1995 Symposium on VLSI Circuits*, pages 115–116, 1995.

28. G. Yoh and F. N. Najm. "A statistical model for electromigration failures". In *2000 IEEE 1st International Conference on Quality Electronic Design (ISQED)*, pages 45–50, San Jose, CA, March 20-22 2000.

29. D. Overhauser, J. R. Lloyd, S. Rochel, G. Steele, and S. Z. Hussain. "Full-chip reliability analysis", Microelectronics Reliability, 38:851–859, 1998.

30. H. Haznedar, M. Gall, V. Zolotov, P.-S. Ku, C. Oh, and R. Panda, "Impact of stress-induced backflow on full-chip electromigration risk assessment", IEEE Transactions on Computer-Aided Design of Integrated Circuits and Systems, 25(6):1038–1046, June 2006.

31. C. Hu, "IC reliability simulation", IEEE J Solid-State Circuits 1992; 27: 241–6. BERT--Circuit Electromigration Simulator. http://techreports.lib.berkeley.edu/accessPages/ERL-90-3.html.

32. J. Warnock. "Circuit design challenges at the 14nm technology node", In *ACM/IEEE 47th Design AutomationConference (DAC-2011)*, pages 464–467, San Diego, CA, June 5-9 2011.

33. Cadence Virtuoso UltraSim Full Chip Simulator Datasheet. Cadence Design Systems. Available from: http://www.cadence.com/.

34. Cadence White paper. Reliability Simulation in Integrated Circuit Design. Available from: http://www.cadence.com/.

35. J.R. Black, "Electromigration - a brief survey and some recent results", IEEE Trans. Elec. Dev., 16 (1969) 338–347.

36. A. H. Fischer, A. Abel, M. Lepper, et al., "Modeling bimodal electromigration failure distributions", Microelectronics Reliability, 41 (2001) 445–53.

37. Ohring M. *Reliability and failure of electronic materials and devices*, San Diego: Academic Press; 1998.

38. J. B. Bernstein, M. Gurfinkel, X. Li, et al., "Electronic circuit reliability modeling", Microelectronics Reliability, 46 (2006) 1957-1979.

39. J.W. McPherson. *Reliability Physics and Engineering*. Springer, New York, NY, 2010.

40. I.A. Blech, C. Herring, "Stress Generation by Electromigration", J. Appl. Phys. Lett. 29 (3) (1976) 131–133.

41. I.A. Blech, "Electromigration in thin aluminum films on titanium nitride", J. Appl. Phys. 47 (4) (1976) 1203–1208.

42. J.R. Lloyd, "New Models for Interconnect Failure in Advanced IC Technology ", *Proc. 14th Int. Symp. On the Physical and Failure Analysis of Integrated Circuits (IPFA), 297 (2008).*

43. V. Sukharev, A. Kteyan, E. Zschech, W.D. Nix, "Microstructure effect on EM-induced degradations in dual inlaid copper interconnects", IEEE TDMR, VOL. 9, (2009) 87-97.

44. M.A. Korhonen, P. Borgesen, K.N. Tu, and Che-Yu Li, "Stress evolution due to electromigration in confined metal lines", J. Appl. Phys., 73 (1993) 3790-3799.

45. J.J. Clement, "Reliability analysis for encapsulated interconnect lines under dc and pulsed dc current using a continuum electromigration transport model", J. Appl. Phys., 82 (1997) 5991-6000.

46. R. Rosenberg, M. Ohring, "Void Formation and Growth During. Electromigration in Thin Films", J. Appl. Phys. 42 (13) (1971) 5671–5679.

47. Z. Suo, "Reliability of Interconnect Structures", in *Interfacial and Nanoscale Failure* (W. Gerberich, W. Yang, Editors), *Comprehensive Structural Integrity* (I. Milne, R.O. Ritchie, B. Karihaloo, Editors-in-Chief), Elsevier, Amsterdam, vol. 8 (2003) 265-324.

48. M.E. Sarychev, Yu.V. Zhitnikov, L. Borucki, et al., "General model for mechanical stress evolution during electromigration", J. Appl. Phys., vol. 86, no. 6 (1999) 3068-3075.

49. H.S. Carslaw and J.C. Jaeger, *Conduction of Heat in Solids*. Clarendon, Oxford, 1947.

50. M. Hauschildt, G. Talut, J. Poppe, et Al, Electromogration void nucleation and growth analysis using large-scale early failure statistics, Book of Abstracts, *12th International Workshop on Stress-Induced Phenomena in Microelectronics*, May 2012, Kyoto, Japan.

51. M. Hauschildt, C. Hennesthal, G. Talut, et Al. "Electromigration Early Failure Void Nucleation and Growth Phenomena in Cu and Cu(Mn) Interconnects", *IEEE Int. Reliab. Phys. Symp.* Proc. (2013) 2C.1.1-6.

52. R.L. de Orio, H. Ceric, S. Selberherr, "A compact model for early electromigration failures of copper dual-damascene interconnects.", Microelectronics and reliability. 51(9-11):1573-1577.

53. A.S. Oates and M.H. Lin, "The Impact of Trench Width and Barrier Thickness on Scaling of the Electromigration Short – Length Effect in Cu / Low-k Interconnects", IEEE Int. Reliab. Phys. Symp. Proc. (2013) 3F.1.1-5.

54. K-B Yeap, M. Roellig, R. Huebner, M. Gall et al, "A critical review on multi-scale material database requirement for accurate three-dimensional IC simulation input", IEEE TDMR 12 (2012) 217-224.

Characterization of Thermal Stresses and Plasticity in Through-Silicon Via Structures for Three-dimensional Integration

Tengfei Jiang[a], Suk-Kyu Ryu[b], Jay Im[a], Rui Huang[b], and Paul S. Ho[a]

[a]Microelectronics Research Center, University of Texas, Austin, TX 78712
[b]Department of Aerospace Engineering and Engineering Mechanics, University of Texas, Austin, TX 78712

Abstract. Through-silicon via (TSV) is a critical element connecting stacked dies in three-dimensional (3D) integration. The mismatch of thermal expansion coefficients between the Cu via and Si can generate significant stresses in the TSV structure to cause reliability problems. In this study, the thermal stress in the TSV structure was measured by the wafer curvature method and its unique stress characteristics were compared to that of a Cu thin film structure. The thermo-mechanical characteristics of the Cu TSV structure were correlated to microstructure evolution during thermal cycling and the local plasticity in Cu in a triaxial stress state. These findings were confirmed by microstructure analysis of the Cu vias and finite element analysis (FEA) of the stress characteristics. In addition, the local plasticity and deformation in and around individual TSVs were measured by synchrotron x-ray microdiffraction to supplement the wafer curvature measurements. The importance and implication of the local plasticity and residual stress on TSV reliabilities are discussed for TSV extrusion and device keep-out zone (KOZ).

Keywords: 3D integration, Through-silicon via (TSV), Thermo-mechanical reliability, FEA.

INTRODUCTION

Three-dimensional (3D) integration with through-silicon vias (TSVs) has emerged as a potential solution to overcome the wiring limit beyond the 22 nm technology node. The TSV is a critical element that provides short vertical interconnects to improve the electrical performance and power consumption for 3D integration [1-4]. This has generated great interests from the industry to develop 3D integration based on TSV technologies. Cu is widely used to form the TSV using a process compatible with Cu backend integration. The fabrication of TSV structures involves deep etching of Si wafer to form via holes, deposition of barrier and seed layers, and the electroplating of Cu to fill the vias. The TSV structures will undergo further processing during the fabrication of 3D integrated circuits, most of which are carried out at elevated temperatures. Thermal stresses can arise during fabrication, testing and operation of the TSV structures due to the large mismatch in the coefficient of thermal expansion (CTE) between Cu and Si. The stresses are large enough to cause serious reliability concerns even structural failures in the integrated structure, including TSV extrusion, cracking of Si near the TSV and degradation of device performance [5-9]. Management and mitigation of thermal stresses in the TSV structures require proper stress characterization and modeling analysis. This is especially challenging since the TSV geometry, material and mechanical behaviors are distinctly different from that of the traditional back end of the line (BEOL) structures.

Experimental methods have been applied to measure thermal stresses in Cu TSV structures. One widely used technique is the micro-Raman spectroscopy, which measures the frequency shift of the Raman modes caused by strain in the Si surrounding the TSVs [10-12]. Depending on the Si orientation and the Raman system configuration, certain stress components for Si or their combinations can be deduced. However, Raman spectroscopy is limited to measuring the near-surface stresses in Si but lacking the ability to analyze the stresses in Cu. Resolving individual stress components can also be difficult and often requires detailed modeling [12]. More recently, synchrotron x-ray microdiffraction has been applied to measure the stress characteristics for TSV structures [13]. X-ray microdiffraction can measure both Si and Cu but so far only very limited results have been reported. In addition, the method requires special facility that is not easily accessible. Numerical analyses such as finite element analysis (FEA) are commonly used to calculate the magnitude and distribution of stresses in TSV structures [14-16]. This method allows the study of complex structures without special experimental facilities. However, proper material and mechanical properties of the constituent materials are required and experimental validation of the model is important.

In this study, we apply the wafer curvature method to study the stress characteristics of Cu TSV structures during thermal cycling. The results are contrasted with those of on-wafer Cu thin film samples for which different curvature behaviors are observed. As a global measurement, the curvature change reflects the overall effect of the thermal stresses in a periodic TSV array, from which the stress behavior of individual Cu TSV can be deduced by numerical analysis. By comparing to thin film samples, we demonstrate that the TSV structures have distinct triaxial stress characteristics as opposed to the biaxial stress state of thin films. This gives rise to residual stresses in the TSVs during thermal cycling with highly localized plasticity near its top at the Cu/Si interface. The stress behavior is correlated to changes in the microstructures of the Cu vias during thermal cycling, which are analyzed by focused ion beam (FIB) and electron backscatter diffraction (EBSD) techniques. Along with the microstructure analysis, the mechanisms underlying the linear and nonlinear temperature-curvature behaviors of the TSV specimen are discussed. The wafer curvature method is supplemented by synchrotron x-ray microdiffraction to measure the local plasticity and deformation in and around individual TSVs. In particular, the local plasticity was observed near the top at the Cu/Si interface, verifying the FEA results. The local plasticity provides a mechanism for TSV pop-up without the presence of interfacial delamination. The implication of the triaxial stress state on TSV reliability is discussed with a focus on the stress-induced carrier mobility change of devices.

CURVATURE MEASUREMENTS DURING THERMAL CYCLING

The TSV test structures used in this study contained patches of TSV arrays fabricated in 780 μm thick (001) silicon wafers. The Cu vias had a diameter of 10 μm and height of 55 μm and with barrier layers of 0.1μm Ta and 0.4μm oxide deposited at the via/Si interface. The spacing between the TSVs was 40 μm along the [110] direction and 50 μm along the [1$\bar{1}$0] direction. After the TSVs were fabricated, annealing was carried out at 100°C for 30 min followed by chemical mechanical planarization (CMP), leaving an oxide layer of 0.8 μm thick on the wafer surface, which was then mechanically removed for the wafer curvature measurements. For the curvature measurement, the wafer was diced into 5 mm by 50 mm strips with each strip containing periodic arrays of blind vias located near the centerline of the sample, as illustrated in Fig. 1. The volume ratio of Cu to Si was 0.015% for the TSV sample. The thin film sample used as a comparison contained 0.8 μm electroplated Cu film on 780 μm Si substrate with a thin diffusion barrier layer in between and a 50 nm SiN capping layer on top of the Cu film. The thin film sample was cut into strips of the same size for the curvature measurement. The volume ratio of Cu to Si in the thin film sample was 0.103%. Since the amount of curvature depends on the volume ratio of Cu to Si, the measured curvatures of the thin film and the TSV are normalized by the Cu volume ratio to facilitate the direct comparison of their stress characteristics.

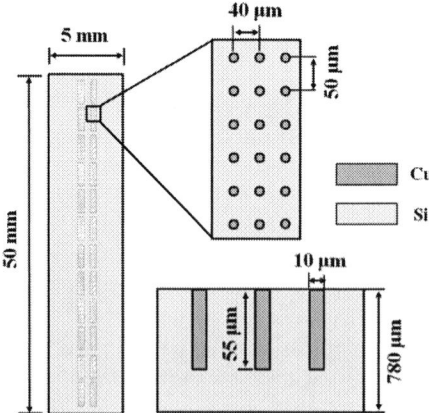

FIGURE 1. Illustration of the TSV sample for the precision wafer curvature measurements.

For the curvature measurements, both samples were subject to six thermal cycles: heating to 200°C and cooling to room temperature (RT) in the first two cycles, followed by two cycles to 350°C and two more cycles to 400°C. After thermal cycling measurements, the same sample was measured again under the same thermal cycling condition with Cu etched off so that a reference curvature, κ_0, can be obtained. By subtracting the reference curvature, the net curvature changes ($\Delta\kappa = \kappa - \kappa_0$) during thermal cycles for the TSV and thin film samples were obtained and plotted in Fig. 2.

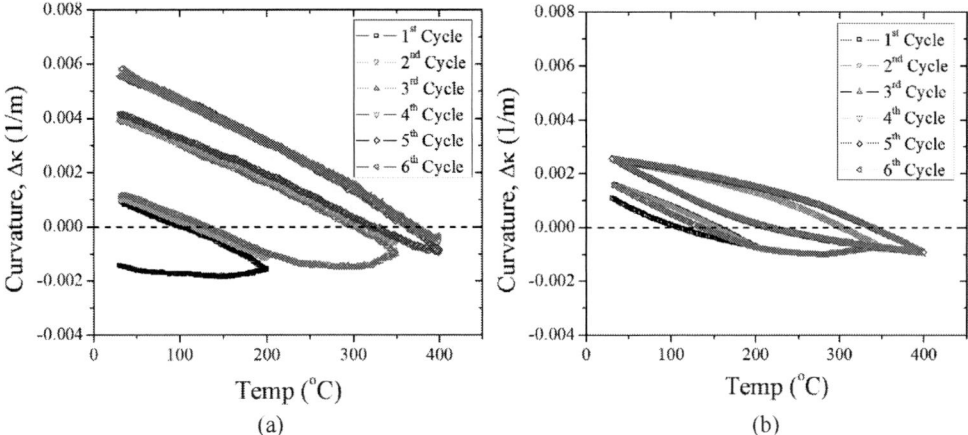

FIGURE 2. Curvature-temperature plots of (a) TSV and (b) electroplated Cu thin film. The curvature of the Cu thin film is normalized by the Cu volume ratio for comparison with the TSV structure.

For the TSV structure, the curvature changed nonlinearly during heating in the first thermal cycle, while during cooling a linear behavior was observed. The heating and cooling curves of the second cycle overlapped with the prior linear cooling curve, where no hysteresis loop was observed. In the third cycle with increased heating to 350°C, the curvature showed a linear behavior until 200°C then resumed nonlinear relaxation to 350°C. During the fourth cycle with heating to 350°C, the entire curvature again followed a linear behavior. A similar behavior was observed for the subsequent two cycles with heating increased to 400°C. The overall behavior was reproducible in that a nonlinear curvature due to relaxation occurred during heating beyond the previous peak temperature, and it is followed by a linear cooling behavior with an increased curvature upon cooling to RT.

The curvature behavior of the TSV structures as a function of temperature is distinctly different from that of the Cu thin film, where hysteresis loops were formed when the sample was thermal cycled, as readily observed in the second, fourth and last cycle in Fig. 2b. For metal thin films, the stress evolution during thermal cycling has been extensively studied with wafer curvature measurements [17-19]. Grain growth, plasticity, and creep are known mechanisms for stress relaxation in Cu thin films to cause nonlinear thermomechanical behaviors, and hysteresis loops are commonly attributed to plastic deformation in the film [19-22]. Detailed studies have been carried out in our laboratory correlating curvature changes with microstructure evolution for electroplated Cu films and the results confirmed that the nonlinear curvature relaxation observed in Fig. 2b can be attributed to such stress relaxation mechanisms [21,22]. Significantly, hysteresis loops were not observed for TSV structures as shown in Fig. 2a, suggesting that large-scale plasticity is not a dominant stress relaxation mechanism. This leads to distinct stress characteristics for the TSV structure, which is analyzed in the following sections.

STRESS ANALYSIS

To analyze the stress characteristics observed for TSV and thin film samples, we focus first on the curvature behavior observed during the cooling cycles where the hysteresis loops were observed only in thin films. 3D FEA models were constructed for both the TSV and thin film structures using a commercial package, ABAQUS (v6.10). For the TSV structure, half of the via was modeled with symmetric boundary conditions in the [110] and [1$\overline{1}$0] directions to simulate the periodic TSV arrays. The material properties used in the model are: Young's modulus, E_{Cu} = 110 GPa, E_{Si}=130 GPa, and E_{oxide} = 70 GPa; Poisson's ratio, ν_{Cu} = 0.35, ν_{Si} = 0.28, and ν_{oxide} = 0.16. The CTEs are α_{Cu} = 17 ppm/°C, α_{Si} = 2.3 ppm/°C and α_{oxide} = 0.55 ppm/°C. The von-Mises stress, which is the effective shear stress driving plastic deformation, was calculated for both structures for a thermal load of ΔT = -270°C, corresponding to 350°C thermal cycling. The results are plotted in Fig. 3.

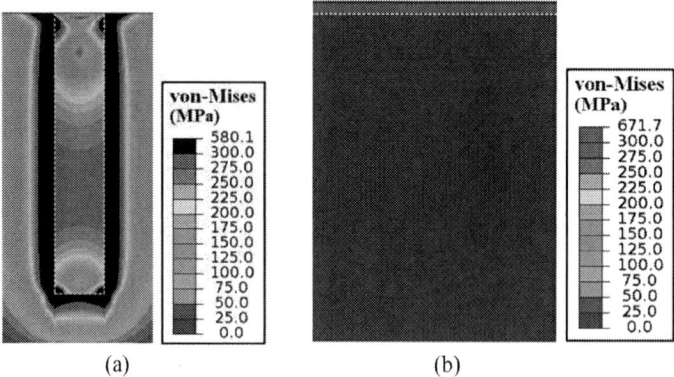

FIGURE 3. Distribution of von-Mises stresses for (a) TSV and (b) thin film structures for a thermal load of ΔT = -270oC, corresponding to 350°C thermal cycling.

For the TSV structure, the von-Mises stress distribution was found to be non-uniform as shown in Fig. 3a. With a thermal load of ΔT = -270°C, the stress level in most part of the via was much lower than the yield strength for Cu, which was reported to range from 150 MPa to 300 MPa, depending on the Cu grain size [23-24]. Thus most of the Cu in the TSV behaves elastically during cooling. This behavior can be attributed to the triaxial stress state in the Cu TSV due to the confinement by the surrounding Si, except for a small region at the top and bottom of the via near the via/Si interface. To demonstrate this behavior, we assumed a relatively high yield strength of 300 MPa and as shown in Fig.3a, with a thermal load of -270°C, most of the via behaved elastically while localized plasticity occurred only near the top and the bottom of the TSV. Since the volume fraction of Cu in the TSV sample is already small, the localized plasticity within Cu vias is too small to affect the overall curvature behavior; therefore hysteresis loop is not observed in Fig. 2a. In contrast, the stress in the Cu thin film is uniform and biaxial due to the two-dimensional confinement of the Cu film by the Si substrate. In this case, the von-Mises stress is uniform and large throughout the film. As a result, the entire film undergoes plastic deformation even for an assumed yield strength as high as 300 MPa, as shown in Fig. 3b. Such large-scale plasticity was manifested by the hysteresis loops observed under thermal cycling to 350°C and 400°C as shown in Fig. 2b. Thus the local plasticity in TSV is a key factor in providing distinct stress characteristics for TSV structures. It plays an important role in controlling the residual stress in the TSV during thermal cycling and has significant implications on stress reliability and device performance. These issues will be addressed in the discussion section.

MICROSTRUCTURE ANALYSIS

Next we analyze the curvature behavior observed during the heating cycle. As shown in Fig. 2, both the thin film and TSV samples followed an initial linear elastic behavior upon heating from the room temperature. The curvature behavior of the thin film specimen became nonlinear at a relatively high temperature, which can be attributed to stress relaxation mechanisms such as plasticity and grain growth and has been well studied [19,22]. For the TSV sample, nonlinear stress relaxation was observed after heating beyond the maximum temperature of the previous heating cycle. As discussed in Section 3, the underlying mechanism for the observed stress relaxation is unlikely due to localized plasticity in very small areas in TSVs. To understand the stress relaxation mechanism for TSV samples, the microstructure evolution was investigated. Several samples were subjected to a single thermal cycle to temperatures ranging from 100°C to 400°C with an interval of 50°C. The curvatures observed are similar to that shown in Fig.2a, as summarized in Fig. 4 for samples cycled to 100, 200, 300, and 400°C.

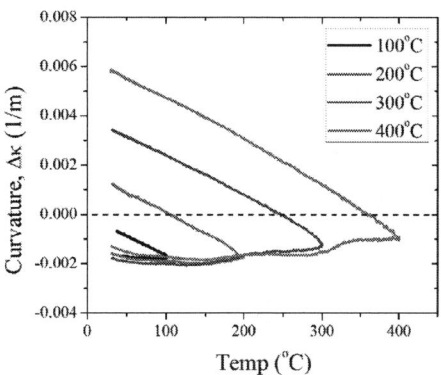

FIGURE 4. Curvature-temperature measurements for TSV samples subjected to one time thermal cycling to temperatures between 100°C and 400°C.

After thermal cycling, the TSV samples were cross-sectioned by FIB for microstructure analysis. Fig. 5 shows the ion channeling images of the via cross-sections after thermal cycling, with the as-received sample also shown as a reference. From the image contrast, an increase in grain size was observed with increasing cycling temperature, providing a qualitative measure of grain growth.

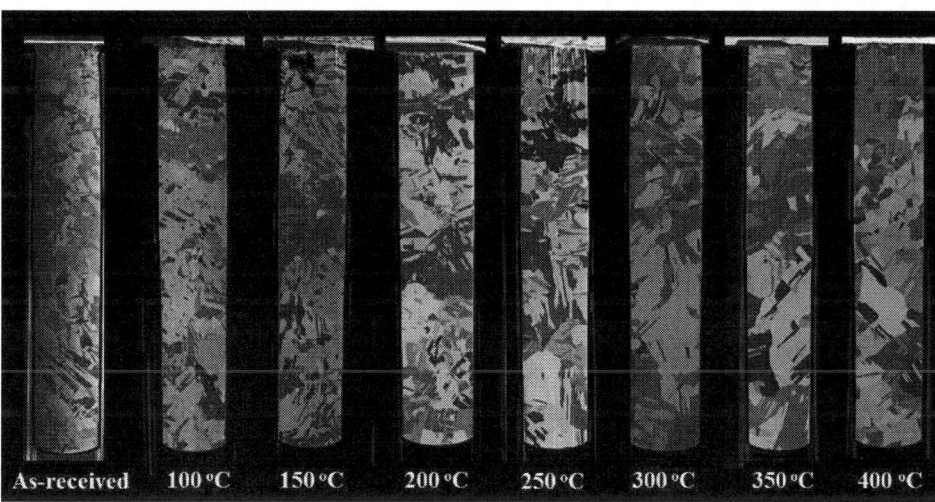

FIGURE 5. Cross-sectional FIB images of TSVs after single temperature thermal cycling measurements. The as-received sample is shown as a reference.

For quantitative analysis of the grain growth of the Cu microstructure, the TSV samples as-received and after thermal cycling were characterized by EBSD. The grain maps of the Cu vias are shown in Fig. 6 as a function of the cycling temperature, with each color corresponding to a particular grain orientation. The average grain size for each sample is measured and plotted in Fig. 6b, clearly showing grain growth with increasing temperature of the thermal cycle.

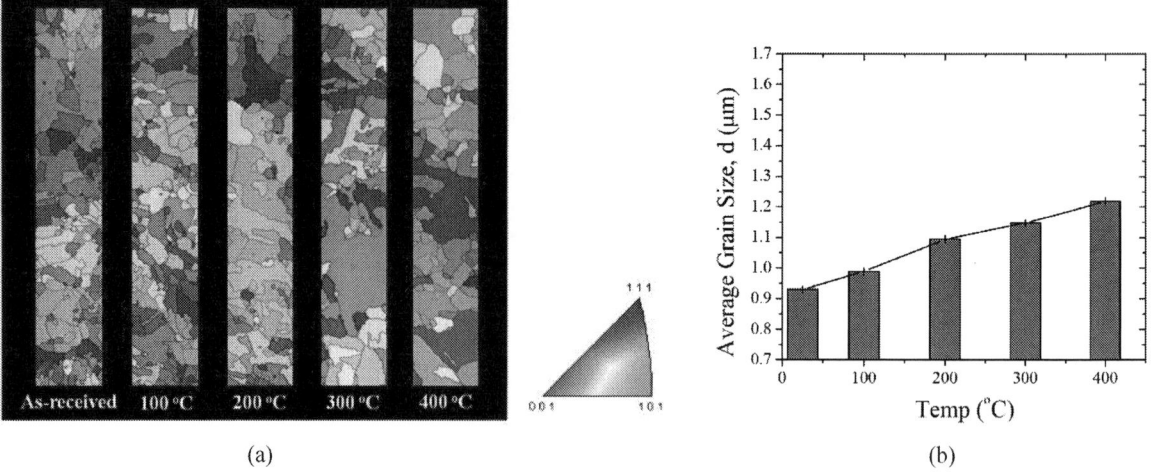

(a)   (b)

FIGURE 6. (a) EBSD grain maps and (b) average Cu grain size for the as-received TSV and TSVs after single temperature thermal cycling.

The grain orientation in the vias was also studied and random grain orientations were found for all samples. Fig. 7 plots the inverse pole figures (IPFs) along the normal direction of the TSV length (ND) for the as-received and thermal cycled samples to 200 and 400°C. The random grain orientation suggests a statistically isotropic Cu microstructure, hence the use of isotropic thermomechanical properties for the Cu vias in numerical analysis is justified.

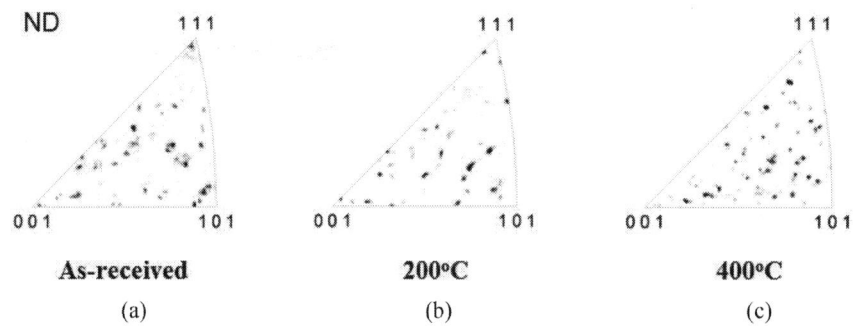

(a)   (b)   (c)

FIGURE 7. Inverse pole figures plotted along the TSV axis (ND) for (a) the as-received TSV, (b) a TSV after thermal cycling to 200°C, and (c) a TSV after thermal cycling to 400°C.

For all samples, a prominent peak was observed at 60° misorientation angles, corresponding to the twin boundaries for Cu. Comparing the statistical distributions of the misorientation angles, the fraction of the twin boundaries was large for the as-received sample (Fig. 8a) and remained large after thermal cycling to 400°C (Fig. 8b). This suggests that twinning was already present after TSV processing. Interestingly, twinning did not seem to contribute to stress relaxation in the TSV structures in this study, although it is an effective mechanism for stress relaxation in Cu thin films [22]. The presence of twin boundaries, however, can increase the Cu yield strength and affect other thermo-mechanical properties of the TSVs [25].

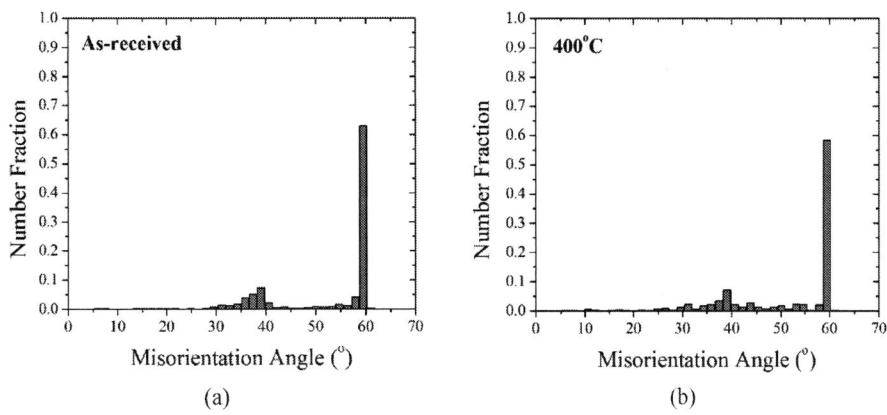

(a)   (b)

FIGURE 8. Grain misorientation angles for (a) as-received TSV and (b) TSV after thermal cycling to 400°C.

DISCUSSION

Stress characteristics of TSVs

Curvature measurements combined with microstructure and stress analyses allow the stress characteristics of TSV structures to be better understood. A key feature of the TSV structure is the triaxial stress state in Cu, which is distinctly different from the biaxial stress state of thin films. Such a difference in the stress states can be traced to the different confinement effects in the 2D and 3D structures by the Si substrate. For the triaxial stress state, the effective shear stress is relatively low in most of the Cu vias so that the TSV behaves mostly linearly elastic during the cooling cycle. Plastic deformation will occur primarily near the top and bottom of the vias and is localized in nature. During the heating cycle, the stress relaxation in the TSV structure is mainly driven by grain growth. By comparing Fig. 2a and Fig. 4, despite the different thermal cycling conditions, the magnitude of the residual curvature appears to be similar when the same peak temperature was reached. This is because grain growth, and thus the stress relaxation, is largely dictated by the peak temperature in the heating cycle [27-28]. In general, it is the stress relaxation due to grain growth during the heating cycle that determines the residual stress in the TSV structure at the end of the cooling cycle.

For quantitative analysis of the residual stress, it is necessary to convert the curvature to stress. For the thin film structure, the curvature can be converted to stress using Stoney's equation, as shown in Fig. 9b [21]. However, for the TSV structure, because of the triaxial stress state and the TSV array configuration, an analytic solution similar to Stoney's equation is not available. In this study, FEA is used to convert the curvature to stress. A 3D model for a quarter of the TSV sample with a symmetric boundary condition is used to simulate the linear curvature behavior observed in the second cycle of the measurement in Fig. 2a. Cu is assumed to be isotropic and linear elastic, and an average curvature is determined along the centerline of the sample for a thermal load of $\Delta T = -200$°C. The rate of curvature change, $\Delta\kappa/\Delta T$, is calculated and compared with the experiment. The calculated $\Delta\kappa/\Delta T = -1.88\times10^{-5}$ m^{-1}/°C is close to the measured curvature change of -1.47×10^{-5} m^{-1}/°C, with the difference possibly caused by the variability of the TSV structures and the uncertainty of material properties. In the experiment, the zero curvature of the sample corresponds to the stress free condition in Cu. This allows a reference temperature to be determined for FEA, and a scaling factor can be obtained by calculating the stress value corresponding to a particular curvature, $\beta = \sigma_{xx}/\Delta\kappa$, which can then be used to convert the measured curvature to stress as shown in Fig. 9a.

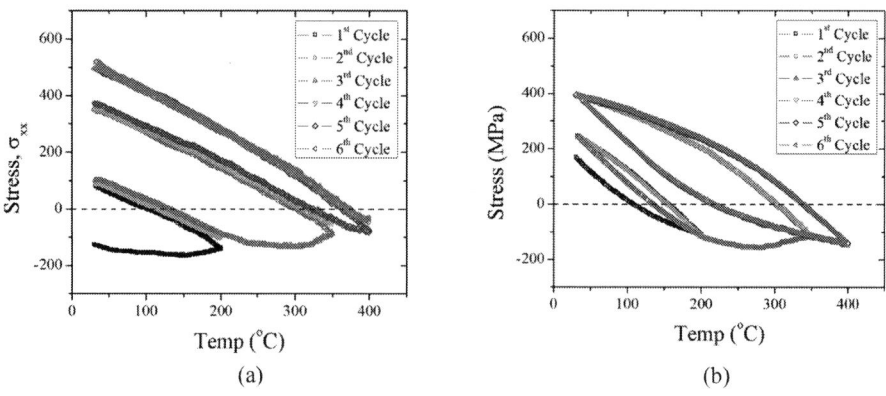

(a)   (b)

FIGURE 9. Conversion of curvature into stress for (a) TSV and (b) thin film structures.

The residual stress in the TSV structure will increases when the TSV is subjected to heating under thermal cycling. For thin films, while stress relaxation by grain growth during heating occurs, the presence of the large-scale plasticity during cooling significantly reduces the residual stress. The effect on the residual stress can be seen by comparing Fig. 9a and 9b, where the residual stress in the TSV is larger than that in the thin film after the 400°C. Given similar amount of stress relaxation during the heating cycles, the residual stresses at the end of the cooling cycles are significantly higher for the TSV. The residual stress has important implications on the device keep-out zone (KOZ) which will be discussed.

Local plasticity and via extrusion

The local plasticity in the Cu via near the top surface and at the Cu/Si interface plays an important role in controlling the via extrusion during thermal cycling. The local plasticity was directly observed by synchrotron x-ray microdiffraction measurements. The x-ray microdiffraction measurement was conducted at beamline 12.3.2 at the Advanced Light Source (ALS), Lawrence Berkeley National Laboratory (LBNL). The cross-sections of an as-received TSV and a TSV thermal cycled to 400°C were scanned with polychromatic x-rays (white beam) with energies from 5keV to 22keV. The x-ray microbeam is focused to a spot size of 1 μm and the scan step size is 1 μm/step. The measurement generates a Laue pattern for each point in the scan matrix which is analyzed with the XMAS software developed at the beamline [29]. The asymmetric broadening of the averaged peak width (APW) can be correlated to increased density of geometrically necessary dislocations in Cu to provide a measure of the local plastic deformation [30, 31]. This is plotted for the Cu via in Fig. 10. Compared to the as-received sample, the Cu via after thermal cycling to 400°C shows a much larger averaged peak width (APW) near the top Cu/Si interface. The increase in the APW and thus the local dislocation density indicates the localized plasticity, while the rest of the via remains nearly elastic. Details of the synchrotron study will be published elsewhere [32].

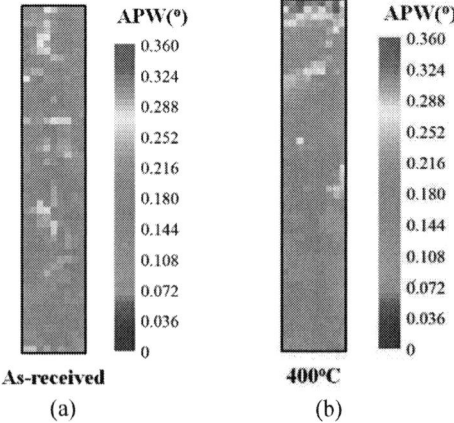

FIGURE 10. Average peak width of Cu via for as-received sample and sample thermal cycled to 400°C obtained by synchrotron x-ray microdiffraction.

The local plasticity has important implication on via extrusion. Via extrusion, or pop-up, is a reliability issue in 3D integration in which Cu extrudes from the wafer surface to damage the interconnect structures above the vias [5]. A previous study has shown that stress-induced interfacial delamination could result in via extrusion [7]. However, in subsequent investigations including this study, via extrusion was observed without evidence of delamination. An alternate mechanism is proposed based on the results from this study that via extrusion is induced by local plasticity in the Cu vias. When the TSV structure is being heated, Cu with the larger CTE will expand but is confined by the surrounding Si. This drives the upward extrusion of Cu above Si. Grain growth will relax the stress, but the stress can still exceed the elastic limit near the top of the via at the interface to induce plastic deformation at an elevated temperature. During cooling, the volume shrinkage of Cu will reduce the amount of the extrusion due to elastic recovery. But as the stress in Cu becomes increasingly tensile, plasticity adds up near the top of the via where the von-Mises stress is largest. The plastic deformation eventually leads to unrecoverable permanent extrusion after the TSV cooled down to room temperature. This process is demonstrated by an elastic-plastic FEA model for a thermal cycle from RT to 350°C. Assuming a yield strength of 250 MPa, the deformation and the equivalent plastic strain of the via at RT, 350°C and after cooling back to RT are shown in Fig. 11. The via extrusion simulated by FEA is consistent with the via extrusion observed in TSV structures [33].

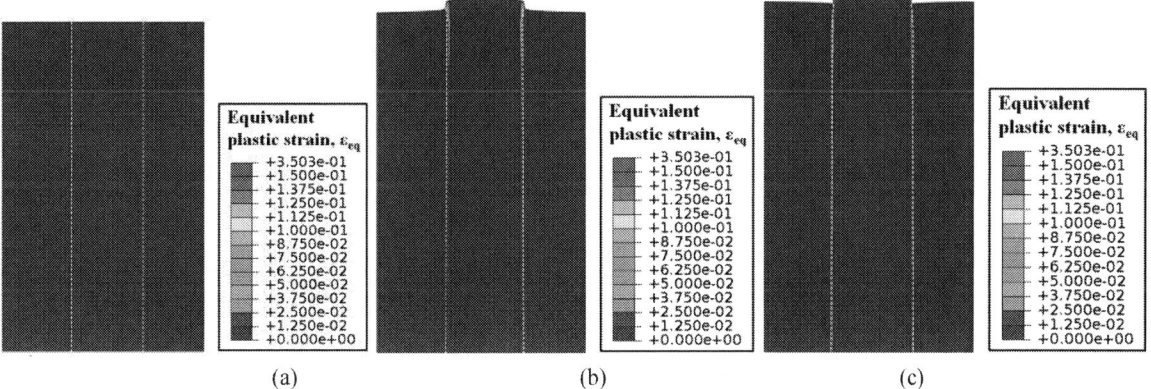

(a)   (b)   (c)

FIGURE 11. Elastic-Plastic FEA model of via extrusion under thermal cycling from RT to 350°C and back to RT with the equivalent plastic strain in Cu plotted. Deformation at different stages of the thermal cycling: (a) at RT before heating, (b) after heating to 350°C and (c) after return to RT. (scale factor =30).

In order to prevent via extrusion, plastic deformation in the TSV structure need to be minimized during the various processing steps carried out at elevated temperatures. This can be made possible through an optimized annealing process, in which the TSV structure is heated first to a maximum temperature T_m (e.g., ~300-350°C) above that for subsequent processing steps, followed by a one-time chemical-mechanical planarization (CMP) process to remove the extruded Cu. Below T_m, grain growth and reduction of Cu yield strength is largely prevented in accordance with the Hall-Petch relation [34]. In addition, strain hardening may accompany plasticity to raise the yield strength locally near the top of the Cu via [21]. With stabilized grain structure and strain hardening, when the TSV is subjected to subsequent processing at temperatures lower than T_m, the plasticity in the via is limited to minimize further via extrusion [5,6].

Residual stress and keep-out zone (KOZ)

The residual stress determined by the curvature measurement has important implications on the stress reliability of the TSV structures, in particular, the keep-out zone (KOZ) for CMOS devices. In modeling the KOZ, a linear elastic model is used as suggested by the linear curvature behavior, and the reference temperature for zero stress is determined from the zero curvature in the measurement. The Cu in TSVs is treated as isotropic. The elastic model used here is justified since the elastic-plastic analysis described in the previous section shows that the plastic deformation is localized and limited to the Si/Cu interface. Based on the measured residual stresses and the elastic behavior, the KOZ surrounding the TSVs and its dependence on the TSV layout were evaluated based on the piezoresistivity effects of Si. Since Si is anisotropic, the KOZ was found to depend on wafer types and the device channel directions. For the [100] channel alignment, a thermal load of $\Delta T = -270$°C resulted in a much larger mobility changes in n-type Si than in p-type Si. With the KOZ defined by a 5% mobility change, KOZ is found to reach 15 um for n-type Si, while the mobility change is less than 5% in p-Si. With [110] channel alignment, the stress effect on mobility is reversed, with large mobility change and KOZ in p-type Si but small mobility change and no KOZ in n-type Si. The local plasticity is found to reduce the amount of mobility change and thus the size of the KOZ. More details of the KOZ study can be found in a separate publication [9].

SUMMARY

A wafer curvature method has been developed and used to study the stress characteristics of Cu TSV structures during thermal cycling. The stress characteristics of Cu TSVs were compared to that of Cu thin films and found to be distinctly different. First and foremost, the stress state in TSV is triaxial, while the stress in thin films is biaxial. For Cu thin films, with a biaxial stress state, the von-Mises stress is often sufficient to cause large-scale plastic yielding in the entire film. In contrast, with a triaxial stress state, the TSV behaves linear elastic with only localized plasticity near the top and bottom at the via/Si interface. Hysteresis loops were observed for thin films due to the large-scale plasticity during thermal cycling, while for the TSV structure, the localized plastic deformation had negligible effect on the overall curvature, and hysteresis loops were not observed. This led to a large residual stress in the Cu TSV at the end of the cooling cycle, making significant impact on the keep-out zone for devices. The residual stress was found to be determined by the stress relaxation occurring during the heating cycle, which is controlled in turn by grain growth in the Cu TSV, depending on the peak temperature in the heating cycle. Since grain growth in Cu is primarily controlled by the electroplating process, the results suggest the possibility to reduce the residual stress in TSV by optimizing Cu electroplating, particularly its chemistry.

Another important result is the finding of local plasticity at the Cu/Si interface near the TSV surface, which can induce via extrusion without interfacial delamination. Synchrotron X-ray microdiffraction was performed and local plasticity was observed in individual TSVs. This provides a new and distinct mechanism for via extrusion. The result suggests that via extrusion can be mitigated by optimizing the microstructure of the Cu TSV to control the local plasticity. The understanding of the local plasticity and the via extrusion mechanism is important for improved reliability of the Cu TSV structures.

ACKNOWLEDGEMENTS

This work was supported by Semiconductor Research Corporation. The authors thank SK Hynix for providing the samples. The authors also thank Drs. Nobumichi Tamura and Martin Kunz at ALS at LBL for helpful discussions. The Advanced Light Source is supported by the Director, Office of Science, Office of Basic Energy Sciences, Materials Sciences Division, of the U.S. Department of Energy under Contract No. DE-AC02-05CH11231 at Lawrence Berkeley National Laboratory and University of California, Berkeley, California. The move of the micro-diffraction program from ALS beamline 7.3.3 onto to the ALS superbend source 12.3.2 was enabled through the NSF grant #0416243.

REFERENCES

1. K. Banerjee, S.J. Souri, P. Kapur, K.C. Saraswat, "3-D ICs: a novel chip design for improving deep-submicrometer interconnect performance and systems-on-chip integration", *Proc. IEEE*, 89(5), pp.602-633 (2001).
2. J.U. Knickerbocker, C.S. Patel, P.S. Andry, C.K. Tsang, L.P. Buchwalter, E.J. Sprogis, G. Hua, R.R. Horton, R.J. Polastre, S.L. Wright, J.M. Cotte, "3-D Silicon Integration and Silicon Packaging Technology Using Silicon Through-Vias", *IEEE J Solid-St Circ,* 41(8), pp.1718- 1725 (2006).
3. J.U. Knickerbocker, P.S. Andry, B. Dang, R.R. Horton, M.J. Interrante, C.S. Patel, R.J. Polastre, K. Sakuma, R. Sirdeshmukh, E.J. Sprogis, S.M. Sri-Jayantha, A.M. Stephens, A.W. Topol, C.K. Tsang, B.C. Webb, S.L. Wright, "Three-dimensional Silicon Integration", *IBM J. Res. & Dev.* 52 (6), pp. 553-569 (2008).
4. J. Lau, "Evolution and Outlook of TSV and 3D IC/Si Integration", El. Packag. Tech. Conf., pp.560-570 (2010).
5. J. Van Olmen, C. Huyghebaert, J. Coenen, J. Van Aelst, E. Sleeckx, A. Van Ammel, S. Armini, G. Katti, J. Vaes, W. Dehaene, E. Beyne, Y. Travaly, "Integration challenges of copper Through Silicon Via (TSV) metallization for 3D-stacked IC integration", *Microelectron Eng.*, 88(5), pp.745-748 (2010).
6. S. Kang, S. Cho, K. Yun, S. Ji, K. Bae, W. Lee, E. Kim, J. Kim, J. Cho, H. Mun, Y.L. Park, "TSV optimization for BEOL interconnection in logic process", *IEEE International 3D Systems Integration Conference (3DIC),* pp.1-3, (2012).

7. S.K. Ryu, K.H. Lu, X. Zhang, J.H. Im, P.S. Ho and R. Huang, "Impact of Near-Surface Thermal Stresses on Interfacial Reliability of Through-Silicon Vias for 3-D Interconnects", *IEEE Trans. Device and Materials Reliability*, 11(1), PP. 35-43 (2011).

8. A. Mercha, G. Van der Plas, V. Moroz, I. De Wolf, P. Asimakopoulos, N. Minas, S. Domae, D. Perry, M. Choi, A. Redolfi, C. Okoro, Y. Yang, J. Van Olmen, S. Thangaraju, D.S. Tezcan, P. Soussan, J.H. Cho, A. Yakovlev, P. Marchal, Y. Travaly, E. Beyne, S. Biesemans, B. Swinnen, "Comprehensive analysis of the impact of single and arrays of through silicon vias induced stress on high-k / metal gate CMOS performance", *Proc. IEEE Electron Device Meeting (IEDM)*, pp. 2.2.1-2.2.4 (2010).

9. S.K. Ryu, K.H. Lu, T. Jiang, J. Im, R. Huang, and P.S. Ho, "Effect of Thermal Stresses on Carrier Mobility and Keep-Out Zone Around Through-Silicon Vias for 3-D Integration", *IEEE Trans. Device and Materials Reliability*, 12(2), pp. 255-262 (2012).

10. I. De Wolf, H.E. Maes, and S.K. Jones, "Stress measurements in silicon devices through Raman spectroscopy: Bridging the gap between theory and experiment", *J. Appl. Phys.*, 79, 7148-7156 (1996).

11. W.S. Kwon, D.T. Alastair, K.H. Teo, S. Gao, T. Ueda, T. Ishigaki, K.T. Kang, and W.S. Yoo, "Stress evolution in surrounding silicon of Cu-filled through-silicon via undergoing thermal annealing by multiwavelength micro-Raman spectroscopy", *Appl. Phys. Lett.*, 98, 232106 (2011).

12. S.K Ryu, Q. Zhao, J. Im, M. Hecker, P.S. Ho, and R. Huang, "Micro-Raman spectroscopy and analysis of near-surface stresses in silicon around through-silicon vias for three-dimensional interconnects", *J. Appl. Phys.*, 111, 063513 (2012).

13. A.S. Budiman, H.-A.-S. Shin, B.-J. Kim, S.-H. Hwang, H.-Y. Son, M.-S. Suh, Q.-H. Chung, K.-Y. Byun, N. Tamura, M. Kunz, Y.-C. Joo, "Measurement of stresses in Cu and Si around through-silicon via by synchrotron X-ray microdiffraction for 3-dimensional integrated circuits", *Microelectron. Reliab.*, 52(3), pp.530-533 (2012).

14. K.H. Lu, X. Zhang, S.K. Ryu, J. Im, R. Huang and P.S. Ho, "Thermo-mechanical reliability of 3-D ICs containing through silicon vias", *Proc. IEEE Electronic Components and Technology Conference (ECTC)*, pp. 630-634 (2009).

15. S.K. Ryu, K.H. Lu, X. Zhang, J.H. Im, P.S. Ho and R. Huang, "Impact of Near-Surface Thermal Stresses on Interfacial Reliability of Through-Silicon Vias for 3-D Interconnects", *IEEE Trans. Device and Materials Reliability*, 11, 35-43 (2011).

16. Moongon Jung, J. Mitra, D.Z. Pan, S.K. Lim, "TSV Stress-Aware Full-Chip Mechanical Reliability Analysis and Optimization for 3-D IC", *IEEE Transactions on Computer-Aided Design of Integrated Circuits and Systems*, 31(8), pp. 1194-1207 (2012).

17. P.A. Flinn, D.S. Gardner, W.D. Nix, "Measurement and Interpretation of stress in aluminum-based metallization as a function of thermal history", *IEEE Transactions on Electron Devices*, 34(3), pp.689-699 (1987).

18. M.D. Thouless, J. Gupta and J.M.E. Harper, "Stress development and relaxation in copper films during thermal cycling", *J. Mater. Res.*, 8(8), pp.1845-1852 (1993).

19. P.A. Flinn, "Measurement and interpretation of stress in copper films as a function of thermal history", *J. Mater. Res.*, 6(7), pp. 1498-1501 (1991).

20. P. Chaudhari, "Grain Growth and Stress Relief in Thin Films", *J. Vac. Sci. Tech.*, 9, 520-522 (1972).

21. D. Gan, P.S. Ho, R. Huang, J. Leu, J. Maiz and T. Scherban, "Isothermal stress relaxation in electroplated Cu films. I. Mass transport measurements", *J. Appl. Phys.*, 97, 103531 (2005).

22. M. Hauschildt, *Effects of Barrier Layer, Annealing and Seed layer Thickness on Microstructure and Thermal Stress in Electroplated Cu Films*, M.S. thesis, University of Texas at Austin, 1999.

23. Y. Xiang, T.Y. Tsui and J.J. Vlassak, "The mechanical properties of freestanding electroplated Cu thin films", *J. Mater. Res.*, 21(6), pp. 1607-1618 (2006).

24. Y. Xiang, J.J. Vlassak, M.T. Perez-Prado, T.Y. Tsui, A.J. McKerrow, "The effects of passivation layer and film thickness on the mechanical behavior of freestanding electroplated Cu thin films with constant microstructure", *Proc. Mater. Res. Soc. Symp.*, 795, pp. 417–422 (2004).

25. L. Lu, Y.F. Shen, X.H. Chen, L.H. Qian, K. Lu, "Ultrahigh Strength and High Electrical Conductivity in Copper", *Science*, 304, 422-426 (2004).

26. D. Gan, P.S. Ho, Y. Pang, R. Huang, J. Leu, J. Maiz and T. Scherban, "Effect of passivation on stress relaxation in electroplated copper films", *J. Mater. Res.*, 21(6), pp. 1512-1518 (2006).

27. G. Dehm, D. Weiss, and E. Arzt, "In situ transmission electron microscopy study of thermal-stress-induced dislocations in a thin Cu film constrained by a Si substrate", *Mater. Sci. Eng. A*, 309-310, pp. 468-472 (2001).

28. M. Lane, R.H. Dauskardt, A. Vainchtein, and H.J. Gao, "Plasticity contributions to interface adhesion in thin-film interconnect structures", *J. Mater. Res.*, 15(12), 2758-2769 (2000).

29. N. Tamura, A.A. MacDowell, R. Spolenak, B.C. Valek, J.C. Bravman, W.L. Brown, R.S. Celestre, H.A. Padmore, B.W. Batterman and J.R. Patel, "Scanning X-ray microdiffraction with submicrometer white beam for strain/stress and orientation mapping in thin films", *J. Synchrotron Rad.*, 10, pp. 137-143 (2003).

30. R. Barabash, G. E. Ice, B. C. Larson, G. M. Pharr, K.-S. Chung, and W. Yang, "White microbeam diffraction from distorted crystals", *Appl. Phys. Lett.*, 79, 749 (2001).

31. B.C. Valek, J.C. Bravman, N. Tamura, A.A. MacDowell, R.S. Celestre, H.A. Padmore, R. Spolenak, W.L. Brown, B.W. Batterman, and J.R. Patel, "Electromigration-induced plastic deformation in passivated metal lines", *Appl. Phys. Lett.*, 81, 4168 (2002).

32. T. Jiang, C. Wu, N. Tamura, M. Kunz, B.G. Kim, H.-Y. Son, M.-S. Suh, J. Im, R. Huang, and P.S. Ho, IEEE Trans. Device and Materials Reliability, DOI: 10.1109/TDMR.2014.2310705.

33. S.K. Ryu, T. Jiang, K.H. Lu, J. Im, H.-Y. Son, K.-Y. Byun, R. Huang, and P.S. Ho, "Characterization of thermal stresses in through-silicon vias for three-dimensional interconnects by bending beam technique", *Appl. Phys. Lett.*, 100, 041901 (2012).
34. D. Hull and D.J. Bacon, *Introduction to Dislocations (4th ed.)*, Butterworth-Heinemann, 2001.

Microstructure, impurity and metal cap effects on Cu electromigration

C.-K. Hu,[a] L.G. Gignac,[a] J. Ohm,[a] C.M. Breslin,[a] E. Huang,[a] G. Bonilla,[a] E. Liniger,[a]
R. Rosenberg,[a] S. Choi,[b] A. H. Simon[b]

[a] IBM T. J. Watson Research Center, Yorktown Heights, NY 10598,
[b] IBM Microelectronic Division, Hopewell Junction, NY 12533

Abstract. Electromigration (EM) lifetimes and void growth of pure Cu, Cu(Mn) alloy, and pure Cu damascene lines with a CoWP cap were measured as a function of grain structure (bamboo, near bamboo, and polycrystalline) and sample temperature. The bamboo grains in a bamboo-polycrystalline grained line play the key role in reducing Cu mass flow. The variation in Cu grain size distribution among the wafers was achieved by varying the metal line height and wafer annealing process step after electroplating Cu and before or after chemical mechanical polishing. The Cu grain size was found to have a large impact on Cu EM lifetime and activation energy, especially for the lines capped with CoWP. The EM activation energy for pure Cu with a CoWP cap from near-bamboo, bamboo-polycrystalline, mostly polycrystalline to polycrystalline only line grain structures was reduced from 2.2 ± 0.2 eV, to 1.7 ± 0.1 eV, to 1.5 ± 0.1 eV, to 0.72 ± 0.05 eV, respectively. The effect of Mn in Cu grain boundary diffusion was found to be dependent on Mn concentration in Cu. The depletion of Cu at the cathode end of the Cu(Mn) line is preceded by an incubation period. Unlike pure Cu lines with void growth at the cathode end and hillocks at the anode end of the line, the hillocks grew at a starting position roughly equal to the Blech critical length from the cathode end of the Cu(Mn) polycrystalline line. The effectiveness of Mn on Cu grain boundary migration can also be qualitatively accounted for by a simple trapping model. The free migration of Cu atoms at grain boundaries is reduced by the presence of Mn due to Cu-solute binding. A large binding energy of 0.5 ± 0.1 eV was observed.

Keywords: Electromigration, microstructure, impurity, CoWP cap, Cu alloy
PACS: 66.30 Qa

INTRODUCTION

Electromigration (EM) mass flow (drift velocity) in Cu damascene lines was determined by Cu effective diffusivity and EM driving force. Once the EM mass flow in a Cu line is determined, then EM lifetime can be estimated. The diffusivity of Cu in Cu lines is related to microstructure and the dominant mass flows are along fast diffusion paths, such as interfaces and grain boundaries (gb). For sub-65 nm technology node Cu lines, the near bamboo-like grain structure observed in previous technology node fine lines no longer exists. A bamboo grain is a single grain spanning the line width and height. The sub-65 nm Cu lines have a mixture of bamboo grains clustered with polycrystalline fine grains. The Cu lines in the sub-65 nm generation nodes also have reduced volumes which further shorten the EM lifetime because there are increased fractions of atoms along the fast paths in conjunction with a smaller void size needed to cause a line to fail. These problems have stimulated great interest among researchers to develop methods to improve Cu EM reliability, including the use of Cu alloys, such as Cu(Al), Cu(Mn), Cu(Mg), Cu(Sn), Cu(Ti) etc., modifying the Cu/dielectric interface, incorporation of a metal capping layer, such as Ta, Ru, Pd, Co, or CoWP, and improvement of the Cu microstructure.[1] The most typical Cu alloy interconnects are fabricated by using a PVD Cu alloy seed layer to replace the pure Cu seed in the line trenches/vias before the Cu electro-chemical plating deposition (ECD) processing step. For the case of a Cu line with a CoWP cap, the top surface of the Cu line was coated with

CoWP from a selective electroless deposition process. Characterization of the mass flow along interfaces and gbs in Cu lines is not only important to gain scientific understanding of EM, but it is also a critical path for predicting and improving Cu reliability. Although EM in Cu alloys has been reported, most of the studies were carried out in samples either with near bamboo-like or bamboo-polycrystalline structures. The existence of bamboo grains in these lines made it difficult to unambiguously distinguish gb and interface contributions separately, since the EM mass flow along interfaces and gbs were strongly coupled in the bamboo-polycrystalline line structures. To separate the contributions, significant differences in Cu grain size distributions between the experimental wafers were achieved by varying the anneal process step after electro-chemical deposition of Cu before or after chemical mechanical polishing. Polycrystalline, bamboo-polycrystalline or near bamboo lines of pure Cu and Cu(Mn) alloys, with or without a CoWP cap, were achieved; thus, the effects of alloying and metal capping, such as with Mn or CoWP, on interface and gb diffusion were able to be resolved separately.

We reported previously that Al enhanced Cu gb diffusion and mitigated Cu/dielectric interface diffusion, and Co had little effect in Cu gb diffusion.[2] The activation energies for Cu alloys using Cu(0.5 at.% Mn) seed were found to be 1.2 ± 0.07 eV, 1.03 ± 0.05 eV, and 0.76 ± 0.05 eV for near bamboo grained, bamboo-polycrystalline, and polycrystalline only lines respectively.[3] The bamboo grains in the bamboo-polycrystalline line act as Cu diffusion blocking boundaries for grain boundary mass flow which generate a mechanical stress induced back flow to compensate the EM grain boundary mass flow: the "Blech short length effect". For this paper, EM in damascene wires with a large variation in grain size and consisting of pure Cu, pure Cu with a thin CoWP surface cap, or Cu(1 - 3 at.% Mn) seed was investigated. The Cu grain size was found to have a large impact on Cu EM lifetime and activation energy, especially for the lines capped with CoWP. We also report that Mn mitigated Cu grain boundary diffusion and was a key to control Cu void growth and hillocks. A simple trapping model was used to obtain the binding energy of Cu and Mn at a Cu gb.

EXPERIMENT

The EM structures used in this work are either two-level M2/V1/M1(EM line)/V1/M2 or three-level polycrystalline Si-NiSi line/WCAvia/M1(EM line)/V1/M2 or M1/V1/ M2 (EM line)/V2/M3, or Blech[4] drift velocity type Cu M1(EM line) segments on W CA lines. Here, M1, M2 and M3 are the first, second and third levels of the Cu interconnect, respectively. The CA, V1 and V2 vias connect NiSi line to M1, M1 lines to M2 lines and M2 lines to M3 lines, respectively. The polycrystalline Si-NiSi and W CA via bars are built in SiN_x/SiO_2 dielectric layers. The Cu M1, M2 and M3 lines and V1 and V2 vias are imbedded in a low dielectric constant material, SiCOH, with a thin amorphous $SiC_xN_yH_z$ cap. The EM Cu line connects to wide and short underlying or overlying lines through vias. EM test lines are 200 μm long and 0.05 to 2 μm wide. A typical damascene level is fabricated by the deposition of a planar dielectric stack, which is then patterned and etched using lithographic and dry-etch techniques to produce the desired wiring or via pattern before depositing the metal. The metals consist of physical vapor deposition

(PVD) TaN/Ta liner, followed by PVD pure Cu or Cu(Mn) seed layer, and then electro-chemically plated (ECD) Cu. The excess metal in the field region is removed using Cu and liner chemical-mechanical polishing (CMP) processes. For type A wafers, a 100°C annealing step was typically used after ECD Cu and before metal CMP so that a large grained Cu damascene microstructure was achieved. The large Cu grain structure resulted from a thick ECD Cu overburden layer and abnormal grain growth behavior in ECD Cu.[5] For the study of microstructure and impurity effects on Cu grain boundary diffusion, fine polycrystalline grained lines were required. Type B wafers did not have the post ECD Cu annealing step and CMP was performed immediately after Cu ECD. The thickness of the metal line and queue time between plated Cu and Cu CMP were found to be critical for obtaining polycrystalline lines.[6] The final Cu lines were passivated with SiO_2/SiN_x, and Al(Cu) metallization was used to coat the bonding pads. The EM test lines were coated with a 25-30 nm thick amorphous $SiC_xN_yH_z$ film cap. To stabilize the final Cu microstructure and impurity distribution, all the wafers were annealed at 400°C for 2h before EM measurements and microstructure studies were performed. Since wafers with large variations in the Cu grain structure were achieved, the effects of Mn impurity and a CoWP surface cap on grain boundary and interface diffusion could be studied separately. The Cu line cross section and microstructure were examined by focused ion beam (FIB), scanning electron microscope (SEM) and transmission electron microscope (TEM) imaging. The thickness and width of the metal lines and vias were measured from TEM images on cross sections prepared by FIB. EM stressing was performed by applying a dc current to the test lines. Most samples were tested in a furnace with air. However, the EM drift velocity samples were tested in a forming gas (N_2-5%H_2) environment furnace and terminated at various times for void growth length study. Failure times, τ, were determined by the length of time required to increase the line resistance from the initial value R_o to $\Delta R/R_o = 1$ to 10%, depending on the failure criteria. Samples were also analyzed for void and hillock location and morphology related to lifetime.

RESULTS and DISCUSION

A. Microstruture

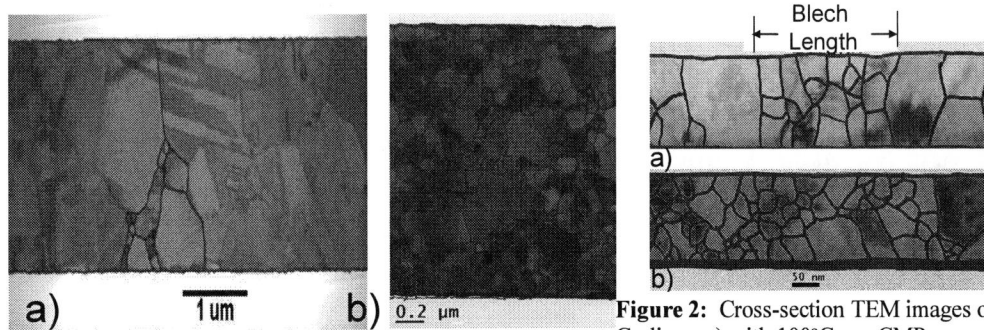

Figure 1: Plan view TEM images of 2 μm wide lines a) 100°C pre-CMP anneal and 400°C post-CMP anneal and b) no pre-CMP anneal and 400°C post-CMP anneal.

Figure 2: Cross-section TEM images of 70 nm wide Cu lines: a) with 100°C pre-CMP anneal and 400°C post process anneal and b) no pre-CMP anneal but with 400°C post process anneal. The "Blech length", the length between two bamboo grains, is also shown.

Figures 1 (a) and (b) are plan view TEM images of 2 μm wide and 0.12 μm thick Cu lines from type A and type B wafers using a pure Cu seed, respectively. Type A wafers had a 100°C thermal annealing step after ECD Cu before CMP. The type B wafers omitted the 100°C thermal annealing step and were CMPed right after ECD Cu. Both types of wafers had a final 400°C post process anneal. In type A wafers, there was a section of large grains with near bamboo grain structure with some visible twins. Fine polycrystalline grains clustered with a few large grains were observed in the type B wafers. The hand-traced lines overlaid on the images are used to illustrate the location of Cu grain boundaries. For the type B wafer, the Cu line height was found to be important to the final grain structure. The grain structure of type B wafer 1.5 µm wide line was multi-grained similar to what is shown in Fig. 1(b). There are a number of large grains but these grains do not span the full line width. Since these films were not annealed prior to CMP to allow grain recrystallization of the plated Cu film, the fine Cu grains formed post ECD were trapped in the lines. Annealing post-CMP will allow grains to grow but the grain size will be limited by the dimension of the 0.12 µm thickness film due to pinning of the grain boundaries by the surface tension of a free surface. A fine grain size, polycrystalline structure with a network of GB paths was observed for 1.5 to 2 µm wide lines with type B processing. This result suggests that wide type B line samples are good candidates for investigating Cu grain boundary diffusion. Fig. 2 (a) and (b) shows TEM cross-sectional images along 70 nm wide Cu lines from type A and B wafers respectively. The TEM images reveal that the 70 nm wide lines from a type A wafer have a mixture of bamboo and polycrystalline grain sections. Sections of multiple small grains are observed through the thickness, but some sections show a single Cu grain extending through the line thickness. In contrast, mostly fine polycrystalline grains regions were seen in the type B wafer. Fig. 2(a) also shows a "Blech Length"[4] which is the length of a polycrystalline line section between two bamboo grains. Since there is low mass flow in bamboo grains especially with a CoWP cap, the EM mass flow in the polycrystalline section induces a stress gradient between two bamboo grains which then will generate a back flow to reduce EM mass flow. [4]

B. EM Mass Flow

The mass flow produced by an EM driving force F_e is given by $J_e = m v_d$, where m and v_d are the atomic density and drift velocity, respectively. The mass flow is expressed as $v_d = m(D_{eff}/k_B T)F_e$, where $F_e = Z^* e \rho j$; e is the absolute value of the electronic charge, Z^* is the apparent effective charge number, ρ is the metallic resistivity, D_{eff} is the effective diffusivity of atoms diffusing through a metal line, T is the absolute temperature, and k_B is the Boltzmann constant. If mass flow along different paths were independent, then the effective diffusivity in a given line at one line cross section could be written as:

$$D_{eff} = \Sigma \, n_i D_i, \hspace{4cm} (1)$$

where the subscript i refers to the ith diffusion path, and n_i and D_i are the fraction of atoms and diffusivity in the ith diffusion path, respectively. Diffusivity is the dominant factor for the mass transport and atoms diffusing along the fast diffusion paths will control the atomic movement. The effective fast diffusion paths in a Cu damascene line are the grain boundaries and the Cu/dielectric interface. For Cu bamboo-like or polycrystalline damascene lines, Cu EM drift velocities for a long line can be written in Eq. (2a) and (2b), respectively, as follows:

$$v_d = [Z_S{}^*(D_S/k_B T)\delta_I(1/h)]\, e\rho j, \tag{2a}$$

$$v_d = [Z_{gb}{}^*(D_{gb}/k_B T)f(\delta_{gb}/d)]\, e\rho j, \tag{2b}$$

where h and d refer to the metal height and grain size, the subscripts S and gb refer to Cu/dielectric cap interface and grain boundary (gb), respectively, δ and D denote the effective width and diffusivity, respectively, and f is a geometric factor. For test structures consisting of mixed bamboo-polycrystalline grain sections in the line, the drift velocity would be

$$v_d = Z_{eff}{}^*(D_{eff}/k_B T)\, e\rho j \tag{3}$$

In this case v_d becomes rather complicated since mass flows along bamboo and polycrystalline grain sections are significantly different and interact strongly. The assumption of independent mass flow along interface and grain boundaries for drift velocity is no longer appropriate. Usually, the slower interface diffusivity will control the net mass flow. The typical EM activation energy, Ea, for interface and gb diffusion of a pure Cu damascene line was shown to be about 1.0 eV and 0.8 eV, respectively,[1] whereas a bamboo Cu line with a CoWP cap was 2.2 eV.[7] The mass flow in a polycrystalline section between two bamboo sections will rapidly change during EM test. The two bamboo grain sections act as mass flow blocking boundaries in the polycrystalline section effectively creating a "Blech short length effect".[4] The fast mass flow in the polycrystalline section will be reduced in time by the compressive stress built-up at the front bamboo-like grain section, although the mass flow or high apparent effective diffusivity in the bamboo-like section is increased by the resultant EM-induced stress gradient from the back and front polycrystalline sections. The drift velocity will be dependent on the fraction and length of the polycrystalline sections and mass flow (flux tunneling)[8] through the bamboo grains. For small polycrystalline section lengths (< Blech critical short length) in Cu lines, the effective mass flow will be controlled by the slow interface mass flow in the bamboo sections. For the case of a long polycrystalline section (>> Blech length) in a Cu line, fast GB diffusion will dominate. The intermediate case becomes an interesting subject to be studied and is not easy to predict since the bamboo grains are not completely blocking boundaries as in the case of the Blech short length effect because of mass leakage through the bamboo grains.

The edge displacement ΔL (void growth) is due to mass depletion at or near the cathode end of the line and $\Delta L_{cr} = \int_0^\tau v_d\, dt$. The lifetime τ can be obtained from

$$\tau \sim \Delta L_{cr}/v_d = \Delta L_{cr}\, k_B T\, /(Z_{eff}{}^* D_{eff}\, e\rho j) \tag{4}$$

τ is the amount of time required to grow a critical void size, ΔL_{cr}, to cause line failure. Since Cu drifts away at the cathode end of the Cu line by the EM driving force, the resistance will rapidly increase when a void grows beyond ΔL_{cr} due to the fact that the high resistivity metal liner in the void has to carry the majority of the current and provides the connection to the remaining Cu line.

C. Effect of Mn in Cu grain boundary diffusion

The EM drift velocity or line damage rate of the polycrystalline Cu lines were greatly increased by adding Al or Mg and reduced with Ti, Sn, Pd, Zr solute additions.[1] These results suggest that the solutes Sn, Pd, Ti and Zr decrease Cu diffusivities in grain boundaries and at free surfaces. In contrast, the impurities Mg and Al were shown to increase Cu grain boundary diffusivity in polycrystalline lines.[9,10] Also no significant mitigation in Cu grain boundary diffusion was observed in low Al, Mn or Co concentration samples using a Cu(1% Al) or Cu(0.5% Mn) seed or a CoWP cap.[2] The addition of Mn or Al solutes was found to cause a reduction in Cu interface diffusion and to increase EM activation energies for Cu-alloy bamboo-polycrystalline lines as compared to pure Cu. However, the effect of Mn concentrations on Cu EM in polycrystalline line structures using Cu > 0.5% Mn seed has not been reported. In this section we will discuss the effect of Mn impurity on Cu grain boundary diffusion using 1.5 μm wide Cu(Mn) polycrystalline line type B samples having a gb network path as shown in Fig 1(b). The polycrystalline damascene lines were fabricated by using a pure Cu or Cu(1 – 3% Mn) seed for ECD Cu, skipping the post-ECD Cu anneal, and performing the CMP step directly after Cu ECD. The Cu alloy line segments were 5, 10, 30, 60, and 100 μm long and on W lines. Mn distribution in the Cu grain boundaries was achieved by a final 400 °C anneal for 2h. These samples were passivated with a 25-30 nm thick $SiC_xH_yN_z$ layer to eliminate Cu free surface diffusion. A dc current density of 50 mA/μm² on the Cu line was applied. The samples were tested as a function of temperature and time for void and hillock growth kinetics study. Void formation was observed for the metal line lengths above 5 μm. This suggested that the critical length for these samples is between 5 and 10 μm.

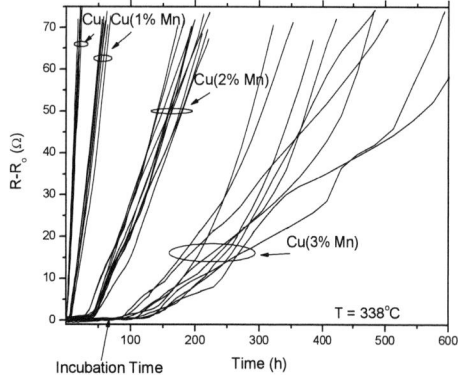

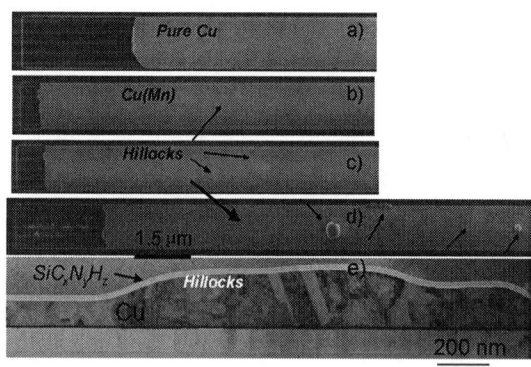

FIGURE 3. The line resistance change vs time for four different Cu metallizations using pure Cu, Cu(1% Mn), Cu(2% Mn) and Cu(3% Mn) seeds. A resistance incubation time, a period with no resistance increase, is also plotted.

FIGURE 4. Top view SEM images of EM tested drift velocity samples near the cathode end of the lines at 338°C a) pure Cu seed and b) to d) Cu(2%Mn) seed with various EM stress times of a) 11h, b) 24 h, c) 48 h, d) 84 h, respectively. e) TEM cross section along the anode end of the pure Cu line. Arrows indicate the hillock locations and electrons flow from left to right.

Figure 3 shows plots of line resistance change as a function of EM stress time for the 1.5 μm wide drift velocity polycrystalline type B samples using pure Cu and Cu(1 to

3% Mn) seeds tested at a sample temperature of 338°C. The plot shows that there is typically a period of small resistance change (incubation period) followed an increase in line resistance. Fig. 4 (a) to (d) are top down SEM images that are used to compare the edge displacement (void length) of pure Cu and Cu(2% Mn) seed lines as a function of time at 338°C. Fig. 4 (e) is a TEM cross section image sectioned along the anode end of a pure Cu line. The diffraction contrast from the Cu grains in the TEM image suggests that the hillock was formed by epitaxial growth of Cu atoms on top of the original Cu line. Fig. 5 shows a plot of void length as a function of time where a slower void growth rate is seen for Cu(Mn) alloy lines than for pure Cu lines. The resistance increase rate and drift velocity are reduced as the Mn concentration is increased. The incubation period is close to zero for pure Cu samples. The incubation period increases as the Mn concentration increases which can be attributed to the fact that Mn is depleted in the cathode end of the line before Cu can be significantly migrated. The hillocks started to form at the critical length (~ 6-8 μm) from the cathode end of the line as shown in Fig. 4(b). The hillocks were grown at the interface between the fast Cu mass flow in the Mn depleted zone and the low mass flow in the Cu(Mn) zone. The movement of the hillock growth location followed the Mn depleted edge interface in the direction of electron flow is shown in Fig. 4 (b) to (d). These observations are similar to the case of EM in Al(2 at.% Cu).[11]

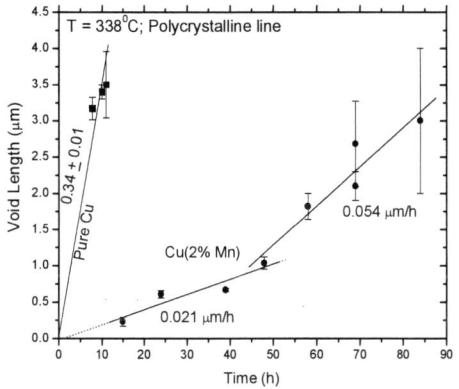

FIGURE 5. Edge displacements of Cu and Cu(Mn) measured at 338°C and 50 mA/μm² as a function of time.

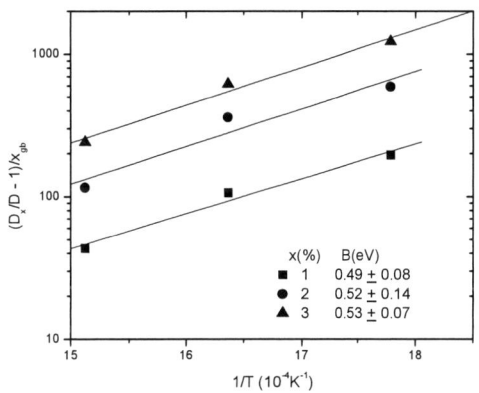

Figure 6: Plot of D/D_x -1 vs. 1/T using polycrystalline only samples. x is the percentage of Mn concentration Cu alloy seed layer.

D. Trapping Model

The reduction of Cu mass flow or diffusivity in Cu(Mn) alloys can be qualitatively accounted for by a trapping model.[12,13] The mobility of Cu at gb is reduced by the presence of Mn such that the free Cu travels rapidly through gb until it meets the Mn where it is trapped. Then one can apply the mass action law to have the relationship between the number of trapped Cu atoms n_t and the number of free Cu atoms n_f :

$$n_t = n_f x_{gb} z \exp(-F/k_B T) \tag{4}$$

where z is the number of degenerate position at the trapping Mn atom, x_{gb} is the Mn concentration in gb and F is the free energy $F = -B - TS$. Here B is the binding energy between solute and Cu vacancy or interstitialcy, which depends on diffusion mechanism in gb, and S is the associated entropy. Assuming the trapped Cu makes no significant contribution to the diffusion, one obtains:

$$[(D/Dx)-1]/x_{gb} = [z \exp(S/k)]\exp(B/k_B T) \qquad (5)$$

The drift velocities of Cu and Cu(Mn) related to diffusivities were obtained by using data for the void growth rate within 1.5 μm void length. The ratios of effective diffusivity D/D_x in Cu gb between pure Cu D and Cu(Mn) alloy D_x were obtained by using drift velocity and Eq. 2(b) with the assumption of similar values of d and $Z^*\rho$ in pure Cu and Cu(Mn) alloys. The left hand side of Eq. (5) is plotted vs 1/T in Fig. 6. Systematically low values of $[(D/Dx)-1]/x_{gb}$ for the lower Mn concentration samples are shown and are presumably due to the estimated errors from the values of x_{gb} which were assumed to be related linearly to the seed layer concentration x % Mn in these calculations. A larger fraction of Mn was piled up at interfaces for the lower Mn concentration. Consequently, the low Mn concentration samples had overestimated x_{gb} values as compared to the high Mn concentration samples. However, a constant and a strong binding energy of 0.5 ± 0.05 eV was obtained from the slopes of these samples. In contrast rather weak Cu vacancy–Mn binding energy in the bulk was observed by positron annihilation technique.[14,15] These results suggest that the mechanistic effects of solute on diffusion in the grain boundaries is fundamentally different than found in the bulk and can be unpredictable, e.g. adding Al solute in Cu, which appears to increase the gb diffusivity.

E. CoWP capping on bamboo-polycrystalline Cu lines

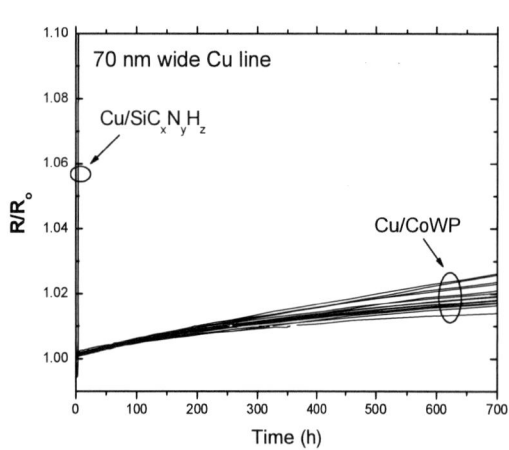

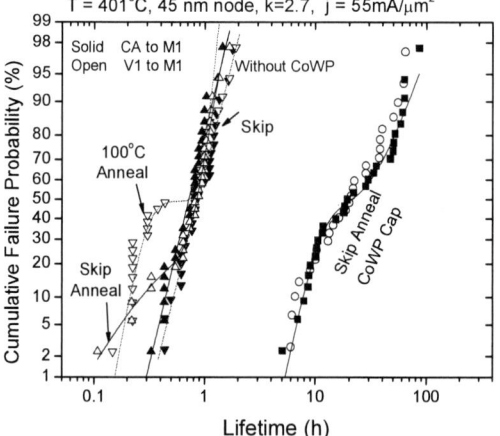

FIGURE 7. Normalized Cu test line resistance as a function of time for lines with and without CoWP cap.

FIGURE 8. Cumulative failure probabilities vs. lifetime for 70 nm wide pure Cu lines with and without 100°C pre-CMP anneal, and pure Cu with a CoWP cap and no pre-CMP anneal.

Fig. 7 shows line resistance curves for 70 nm wide type A M1 lines as a function of time at T = 350°C. These samples have a mixed bamboo-polycrystalline microstructure as shown in Fig. 2 (a). The Cu lines with a CoWP cap had more than 100 x improvements in Cu lifetime as compared to samples without a CoWP cap. The initial line resistance drop before a sudden resistance increase for pure Cu lines without a CoWP cap is due to EM impurity purification for plated Cu and/or a contact resistance reduction. No line damage in samples with CoWP were observed even after 730 h of testing. A small line resistance increase for Cu lines with a CoWP cap was due to the diffusion and low solubility of Co in Cu.[7] In Fig. 8, the EM lifetime distributions of 70 nm wide pure Cu lines (type B) with a CoWP cap, and type A and B lines without CoWP cap at the sample temperature of 401°C are shown. For the case of 1.5 to 2 µm wide lines (type B), a polycrystalline line structure was observed. Conversely, 70 nm wide type B lines have some bamboo grain sections because 70 nm wide Cu lines have a higher probability of obtaining bamboo grains than 1.5 µm wide lines. The data also showed a similar EM lifetime for type A and B wafers without a CoWP cap. Previously, single log-normal and bimodal log-normal distribution functions were observed in the pure Cu without CoWP for Cu mass flow from W CA to Cu M1 and Cu V1 to Cu M1, respectively. This result indicated that the cathode end of line/via contact structure influenced the failure distribution.[1] The resistance change is sensitive to the void locations in the test structure; for example, a contact interface void or line void for the V1 to M1 case and a line void for CA to M1. The bimodal distribution function with similar lifetimes obtained for type B grain structure lines with a CoWP cap indicated that the cathode end of the line contact did not have a critical effect on lifetime and that the weak Blech length links in the M1 line were most likely causing the failure. Fig. 8 still shows a huge lifetime enhancement suggesting that the 70 nm wide fine grain lines were not completely polycrystalline and had some bamboo grains.

Figures 9 and 10 show TEM images of the EM tested type B and type A 50 nm wide lines with a CoWP cap, respectively. A void at the Cu line/W CA interface was seen in Fig. 9 with electron flow from W CA to Cu M1, although typically a void was not often observed. For the case of mass flow from V1 to M1, voids under Cu V1 via and on top of the Cu line surface under CoWP layer were observed (Fig. 10). There are also many voids on top of small and larger Cu grain surfaces under CoWP capped layer in this sample (not shown).

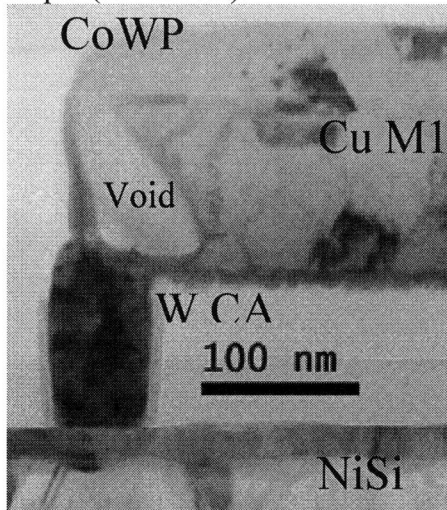

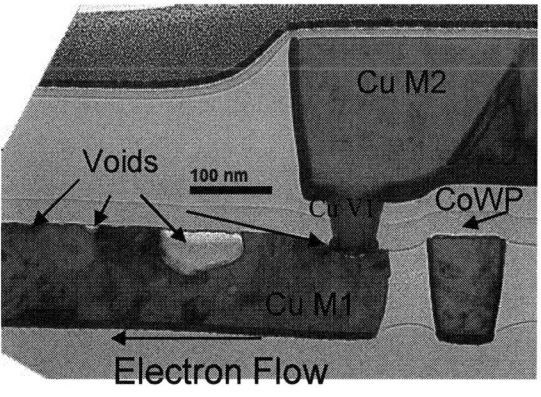

FIGURE 9. Cross-section TEM image of type B wafer at the cathode end of 50 nm wide M1 line. Void location was at the bottom of the M1 line at the W CA interface.

FIGURE 10. Cross-section TEM image of EM tested type A wafer Cu V1 to 50 nm wide M1 line with a CoWP cap. Void locations were under the bottom of the V1 and along the top surface of the Cu line.

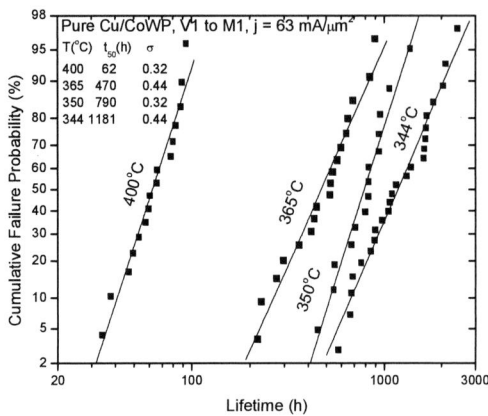
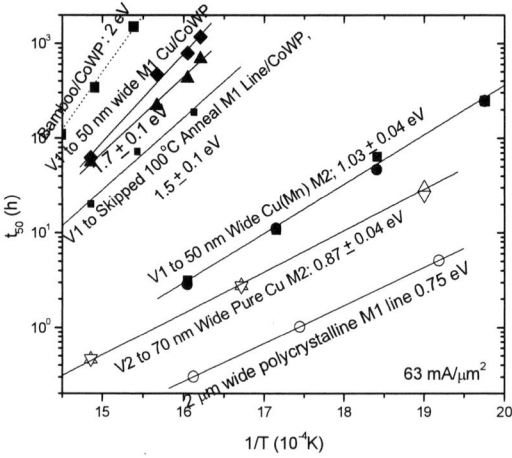

FIGURE 11. Plot of cumulative lifetime probability vs lifetime for 50 nm wide type A lines with a CoWP cap for various sample temperatures.

FIGURE 12. Plot of t50 vs 1/T for various samples: pure Cu, Cu (Mn) alloy, pure Cu with a CoWP cap with type A and B grain structures.

Figure 11 shows the cumulative lifetime distribution for 50 nm wide type A lines with a CoWP cap as a function of sample temperature using a V1 to M1 test structure. Single log-normal distributions are shown with reasonably good σ values of 0.3–0.5. The measured median lifetimes as a function of 1/T are plotted in Fig. 12 for various wafer types and line widths together with data from Ref. [3, 7]. Models which assume gb and interface mass flows to be independent resulting in the equations for lifetime $\tau = \Delta L_{cr}/v_d$ either $\tau = \Delta L_{cr}/[p_{gb}v_{gb}+(1-p_{gb})v_s] \sim \Delta L_{cr}/(p_{gb}v_{gb})$ or $\tau = p_{gb}\Delta L_{cr}/v_{gb}+ (1-p_{gb}) \Delta L_{cr}/v_s \sim (1-p_{gb}) \Delta L_{cr}/v_s$ would not fit the experimental data on bamboo-polycrystalline lines since both equations ignore the interlock effect between interface and gb mass flows. p_{gb} is the fraction of polycrystalline grain sections in the line. Thus the empirical equation $\tau = \Delta L_{cr}/v_d = \tau_o \exp(E_a/k_bT)$ was used. The values of EM activation energy E_a for these Cu lines with CoWP were measured to be about 1.5 eV and 1.7 eV for 70 nm wide lines with a fine type B grain structure and 50 nm wide type A lines, respectively. The 70 nm wide bamboo-polycrystalline pure Cu line (type A) without a CoWP cap had an Ea of 0.87 eV. For comparison, 50 nm wide Cu(Mn) alloy (type A) data points were also included.[3] The data in the figure illustrate that change in the Cu grain structure can effectively produce a remarkable change in overall Cu Ea and lifetime. Similar observations of grain size effect on lifetime for Cu lines with a metal capping layer have been reported.[16] The enhanced Cu lifetimes in Cu lines with a CoWP cap were obtained in 70 nm wide type B lines, 50 nm wide bamboo-polycrystalline type A lines, and 0.2 μm wide near-bamboo type A lines but not in 1.5 to 2 μm wide polycrystalline[1] only type B lines. These results suggest that the existence of bamboo grain sections in lines play a key role in slowing down void growth rate as discussed previously. The enhancements were believed to be due to the bamboo grain sections in the bamboo-polycrystalline Cu line acting as Cu diffusion blocking boundaries for GB mass flow and generating a mechanically induced back flow to reduce the EM GB mass flow. This mechanism is similar to the Blech short length effect which can prolong Cu lifetime. Bamboo grains were observed using cross-sectional TEM imaging along the direction of the 50 nm or 70 nm wide lines over a dimension of a few microns. The polycrystalline length section is shorter than the estimated Blech critical length of 5 μm in Cu. Further study of EM behaviors in Cu structure with varying polycrystalline line length and fraction of bamboo line section remains an important and interesting topic.

CONCLUSIONS

This paper describes the effects of microstructure, impurity and CoWP cap on EM for Cu damascene interconnects. The investigated grain structures in Cu interconnect ranged from near-bamboo, to polycrystalline mixed with single crystal bamboo-like grains, to polycrystalline-only fine grains. The Cu grain microstructures were examined by SEM, FIB, or TEM. The variation in microstructure for Cu lines was achieved by grain growth pinning by the line size effect and by adjusting the thermal annealing steps post-Cu plating before and after CMP. The mass transport in Cu interconnects occurs mainly by interface and/or grain boundary diffusions which depend on Cu microstructure.

The grain size and length of polycrystalline sections in Cu lines with a CoWP cap strongly affect the EM lifetime, and *Ea* was reduced from 2.2 eV, 1.7 eV, 1.5 eV, to 0.72 eV for near bamboo, bamboo-polycrystalline, most polycrystalline, and polycrystalline only lines, respectively. The effect of Mn in Cu grain boundary diffusion was found to be dependent on Mn concentrations in Cu. For the case of Cu(1-3% Mn) seed samples, the depletion of Cu at the cathode end of the Cu(Mn) polycrystalline line was preceded by an incubation period in which Mn was depleted. Unlike pure Cu line with void growth at the cathode end and hillocks at the anode end of the line, the hillocks grew at a starting position about equal to the Blech critical length away from the cathode end of the Cu(Mn) line. The effect of Mn on Cu grain boundaries migration can also be qualitatively accounted for from a simple trapping model. The free migration of Cu at grain boundaries is reduced by the presence of Mn due to Cu-solute binding. A large binding energy of 0.5 eV was observed.

The observed Cu lifetime enhancements for Cu(1%Al)-seed, Cu(0.5% Mn)-seed lines or pure Cu line with a CoWP cap are believed to be due to the presence of bamboo grains in the bamboo-polycrystalline line structures. The average bamboo grain separation is < 1 μm along 70 nm wide bamboo-polycrystalline lines from samples investigated by TEM analyses. It is proposed that Co, Mn or Al at bamboo grain surface acts as Cu diffusion blocking boundaries for Cu gb mass flow, resulting in a similar effect to the "Blech Short Length". High concentration of Mn can produce lines with very long EM lifetimes but there is a trade-off between reliability and performance. Further techniques for improving Cu nanowire microstructure to increase the number of bamboo-like grain sections and reduce the impurity effect on Cu resistivity and EM performance could be critical for the development of 14 nm node and beyond interconnect technologies.

ACKNOWLEDGMENTS

We would like to acknowledge the efforts of our colleagues at the IBM Microelectronics Division, East Fishkill Facility, Hopewell Junction, NY, for the fabrication of samples.   Part of the work for this paper was performed by Research Alliance Teams at various IBM Research and Development Facilities

References

1. C-K. Hu, R. Huebner and P. S. Ho, in "Advanced Interconnects for ULSI Technology", M. R. Baklanov, P.S. Ho and E. Zschsch editor (John Wiley & Sons, Ltd. of The Atrium, West Sussex, 2012) Chap. 9
2. C.-K. Hu, M. Angyal, B.C. Baker, G. Bonilla, C. Cabral, D. F. Canaperi, S. Choi, L. Clevenger, D. Edelstein, L. Gignac, E. Huang, J. Kelly, B. Y. Kim, V. Kyei-Fordjour, S. L. Manikonda, J. Maniscalco, S. Mittal, T. Nogami, C. Parks, R. Rosenberg, A. Simon, Y. Xu, T.A. Vo, C. Witt, AIP Conf. Proc. 1300 (2010) 57
3. C.-K. Hu, J. Ohm, L. M. Gignac, C. M. Breslin, S. Mittal, G. Bonilla, D. Edelstein, R. Rosenberg, S. Choi, J. J. An, A. H. Simon, M. S. Angyal, L. Clevenger, J. Maniscalco, T. Nogami, C. Penny, J. Appl. Phys. 111, 093722 (2012)
4. A. Blech, *J. Appl. Phys.,* 47, 1203-1208 (1976)
5. J.M.E. Harper, C. Cabral Jr., P. Andricacos, L.M. Gignac, I. Noyan, K. Rodbell and C.-K. Hu, J. Appl. Phys., 86, 2516 (1999)
6. A. K. Stamper,H. Baks,E.Cooney,L. Gignac,J. Gill, C-K. Hu,T. Kane,E. Liniger, Y-Y. Wang, and J. Wynne, Proc. of Advanced Metallization Conf. (Mat. Res. Soc., Warrendale, PA 2005) p.727
7. C.-K. Hu, L. M. Gignac, R. Rosenberg, B. Herbst, S. Smith, J. Rubino, D. Canaperi, S. T. Chen, S. C. Seo, and D. Restaino, Appl. Phys. Lett. 84, 4986 (2004).
8. M.A. Korhonen, T. Liu, D.D. Brown and C.-Y. Li, AIP Conf. Proc. , 373, 117 (1995)
9. C.-K. Hu, K.Y. Lee, K.L. Lee, C. Crabal, E. Colgan, C. Stanis, *J. Electrochem. Soc.,* 143, 1001-1006 (1996).
10. N.L. Michael and C.U. Kim, *J. Appl. Phys.,* 90, 4370-4377 (2001).
11. C-K. Hu, P.S. Ho and M.B. Small, J. Appl. Phys. 72,291 (1992)
12. C-K. Hu and H.B. Huntington, in Diffusion Phenomena in Thin Films, edited by G. Gupta, P.S. Ho (Noyes Publications 1988) Chap. 10
13. R. Rosenberg, *J. Vac. Sci. Technol.*, 9:263 (1972)
14. H. Fukushima and M. Doyama, J. Phys. F: Metal Phys., 6, 677 (1976)
15. U. Klemradt, B. Drittler, T. Hoshino, R. Zeller, P.H. Dederichs, and N. Stefanou, Phys. Rev. B, 43, 9487 (1991)
16. L. Zhang, M. Kraatz, O. Aubel, C. Hennesthal, E. Zschech and P.S. Ho, AIP Proc. 1300 (Melville, NY, 2010) p.3

Advanced Concepts for TDDB Reliability in Conjunction with 3D Stress

Martin Gall, Kong Boon Yeap, and Ehrenfried Zschech

Fraunhofer Institute for Ceramic Technologies and Systems (IKTS)
Maria-Reiche-Str. 2, D-01109 Dresden, Germany

Abstract. Time-Dependent Dielectric Breakdown (TDDB) in the Backend-of-Line (BEoL) stack has become one of the most important failure mechanisms for state-of-the art integrated circuits and threatens the long-term reliability of advanced semiconductor products. The continuous reduction in the BEoL feature sizes, resulting in continuously smaller spacing between interconnects, along with a slower pace in the reduction of the operational voltages, has led to significantly increased electrical fields. In addition, the introduction of low-k and ultra-low-k (ULK) materials has complicated the situation even more, since lifetimes for those materials are typically several decades shorter than for traditionally used SiO_2. While the reliability community has mostly adopted the square-root-E model for backend dielectric failure and has abandoned the more conservative linear-E-model, major questions about the true physical mechanism of dielectric failure, such as the role of Cu, still exist. Within the context of "More than Moore", the 3D IC integration approach promises to give a significant performance boost, power savings and cost reduction. However, globally and locally induced stresses, as well as additional complexities and interactions caused by the 3D process will influence the physical failure mechanisms for TDDB, leading to even shorter lifetimes. Possibly, lifetime data acquired with standard testing and extrapolation methods may overestimate the actual product lifetimes. Therefore, advanced concepts for the evaluation of BEoL "dielectric reliability" have to be developed. In this paper, we review the current status of TDDB testing, address the main issues and open questions, and finally we propose new concepts for a more realistic evaluation of the "dielectric reliability" within the context of mechanical stresses caused by 3D IC integration schemes.

Keywords: Time-Dependent Dielectric Breakdown, BEoL, Failure Mechanisms, Dielectric Reliability, 3D IC Integration, Mechanical Stress, Diffusivity

INTRODUCTION

Prior to approximately the year 2000, generating a fully integrated BEoL interconnect process flow for advanced semiconductor technology was viewed as an evolutionary effort, because much of the work in advanced CMOS integration focused on meeting only the technological demands of dimensional scaling [1]. In essence, the critical dimensions continued to scale downward by a factor of about 0.7 per technology generation, but the materials and processes that were used to build up the integrated stack were essentially the same, namely Al (with Cu alloying) for the metallization and SiO_2 as the inter-layer dielectrics. This situation, however, changed towards the end of the 1990s after it became apparent that new BEoL materials would be needed. The introduction of new materials (Cu to reduced electrical resistivance and to improve reliability, and low-k dielectrics to reduce capacitance) was needed to ameliorate the effects of increased RC interconnect delay due to interconnect dimensional shrinkage [2, 3]. Thus, dimensional scaling still progressed at roughly the same rate as before, but new materials were introduced at a sub-250 nm critical (interconnect) dimension to provide an additional performance benefit to dimensional scaling. Firstly, Cu metallization was introduced into manufacturing in 1997 [4, 5] and, subsequently, low-k dielectrics of different types were incorporated into the BEoL stack [e.g., 6, 7], starting with the incorporation of F-doped silica dielectrics at the 180 nm CMOS technology node. The introduction of these new materials, however, significanly increased the efforts to retain the benefits of interconnect performance scaling from process, integration, and reliability perspectives. Since Cu (with some alloying, depending on the application) is likely to remain the interconnect metallization of choice for the foreseeable future, the main challenge for new materials integration has shifted toward the implementation of progressively lower-k dielectrics. In very advanced CMOS technology nodes, ULK dielectrics are being introduced, mostly based on various types of organosilicate glass (SiCOH-type) films with varying amounts of porosity. Figure 1 shows the normalized electrical breakdown fields for various dielectrics, starting with the initially used, mechanically and electrically strong Tetraethyl Orthosilicate (TEOS)-based SiO_2 films which were already implemented in Al(Cu)-based technologies decades ago. The strength decreases by about 30% when switching towards dense SiCOH-type materials, and about another 30% for ULK materials with a k-value of about 2.3, in this case a Methylsilsesquioxane (MSQ) film [8]. Together with the dimensional down-scaling and a slower pace in the reduction of the operational voltages, this leads to a significantly increased risk due to TDDB failures.

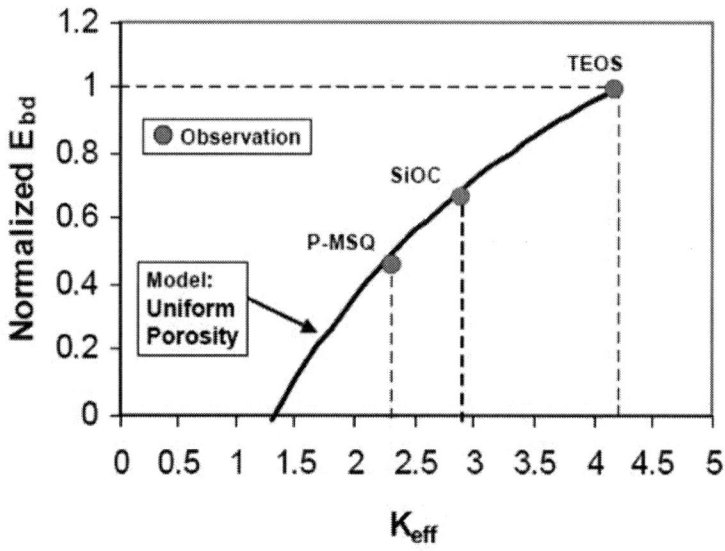

FIGURE 1. Relative electrical breakdown strengths for various dielectric materials showing the significantly reduced E_{bd} values for ULK films [8].

TDDB FAILURE MODES AND MECHANISMS

Figure 2 shows two possible failure modes for TDDB to occur under the assumption that Cu diffusion through the dielectrics is one of the major factors controlling the lifetime. The scheme shows two interconnect structures which are biased with respect to each other. Typically, it is assumed that Cu is positively ionized, and therefore, Cu atoms migrate from the positively charged side to the negatively charged side, following the electric field direction. In the case on the left of Fig. 2, diffusion along the dielectric/cap interface constitutes the failure mechanism. The cap usually consists of a SiN or SiCN film. In the case on the right of Fig. 2, bulk diffusion through the dielectric leads to the TDDB failure. Due to the generally well-controlled, high-quality barrier layers between the Cu and the dielectrics, typically a bilayer combination of TaN and Ta, it is mostly assumed that interface diffusion controls the lifetime. In addition, the process of chemical-mechanical polishing (CMP) can damage the surface of the dielectrics. Even with plasma-assisted cleaning before the SiN/SiCN deposition, one can assume that the interface is the weakest spot in the BEoL stack. Figure 3 shows a typical transmission electron microscope (TEM) image of a TDDB failure [9]. A "cloud" of Cu can be seen between the interconnect structures, most likely originating from the top corner of the positively charged side to the left of the test structure.

Interface diffusion   "Bulk" dielectric diffusion

FIGURE 2. Two possible failure mechanisms for TDDB to occur under the assumption that Cu diffusion through the dielectrics is one of the major factors controlling the lifetime [9].

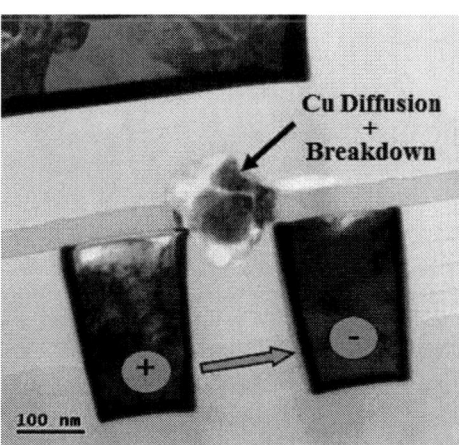

FIGURE 3. Typical TEM image of a TDDB failure [9].

Figure 4 depicts the proposed failure mechanism under the assumption that Cu diffusion is involved [10]. The underlying model proposes an electron fluence-driven, Cu-catalyzed SiCOH breakdown mechanism, and it provides the basis for the "square-root-E" dependence often seen for SiCOH TDDB analysis.

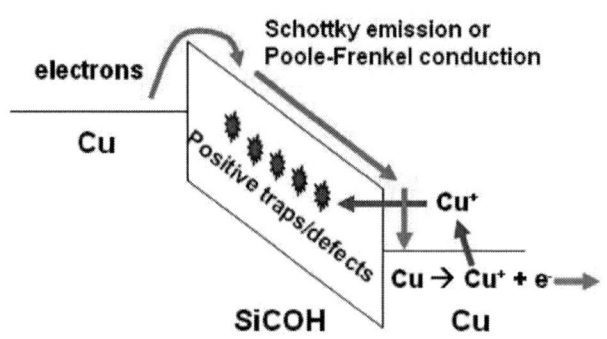

FIGURE 4. Proposed TDDB failure mechanism involving Cu diffusion [10].

It is postulated that accelerated electrons injected from the cathode follow a Schottky emission or Poole-Frenkel conduction mechanism during their transport inside the SiCOH dielectrics. Some of the "lucky" electrons could undergo "thermalization" under high field and high temperature conditions, and a fraction of such energetic electrons could impact the Cu atoms at the anode and accelerate the generation of positive Cu ions when they reach the anode. Those generated Cu ions could then inject into the dielectrics under the field along a fast diffusion path (such as the SiCOH/cap interface) to create damage in the SiCOH film. Migrated Cu ions could recombine with electrons to become Cu atoms. Two possibilities of SiCOH breakdown could happen after the concentration of Cu in SiCOH reaches a critical level. Accumulated Cu atoms in the s could form clusters or nano-particles, and such agglomerates could be connected eventually, to create a direct metallic shorting bridge or to cause effective local dielectrics thinning to trigger an electrical short. Alternatively, the diffused Cu atoms can catalyze the bond breakage reaction of SiCOH easily because of their relatively large atom size, which is large enough to induce permanent SiCOH bond displacement either by interstitial insertion of a local strain field or by direct atom collision. The Schottky emission or Poole-Frenkel conduction mechanism, as mentioned above, leads to the following dependence of the TDDB median time to fail (*MTTF*) on the applied electrical field E and temperature T:

$$ MTTF = A \cdot \exp(-\gamma(T) \cdot \sqrt{E}) \cdot \exp\left(\frac{E_A}{k_B \cdot T} \right) \qquad (1) $$

where A is a constant, $\gamma(T)$ the temperature-dependent field acceleration parameter, E_a the activation energy and $k_b T$ the thermal energy [1]. The Schottky emission or Poole-Frenkel conduction mechanism provides the basis for this so-called "square-root-E-model" which has widely replaced the previously used "linear-E-model". An alternate approach, not necessarily involving Cu diffusion, is the so-called "impact model" [11, 12], considering an electron being accelerated in a dielectrics caused by an electric field (Fig. 5). This electron will most likely be scattered somewhere in the dielectrics, resulting in a mean free path for this event to occur. One can make the assumption that the electron will be gaining energy from the field until it is scattered, whereupon the energy that has been contributed by the field to the electron is transmitted to the scattering site. If the energy is sufficiently high, it will create a defect or a "trap" that will contribute to the final breakdown of the dielectrics.

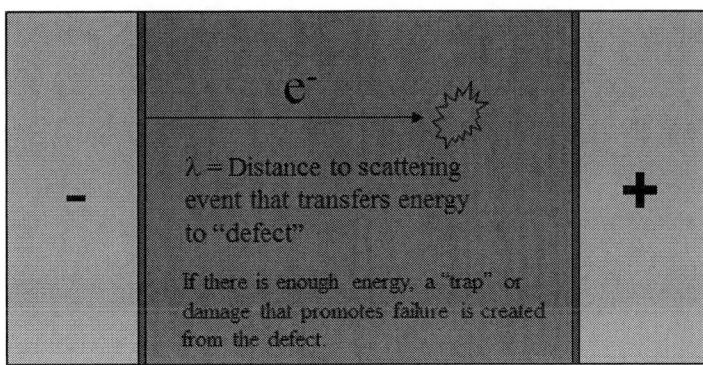

FIGURE 5. Proposed TDDB failure mechanism not involving Cu diffusion [11, 12].

It is reasonable to expect that the time to breakdown would be proportional to the rate at which damage is created, which, in turn, is proportional to the product of the leakage current and the probability that enough energy was available at the time of the collision to create the damage. Given that an electron will gain energy as long as it is accelerated in the electric field, the probability that there will be a damaging collision is the probability that an electron, given a mean free path, will have traveled a sufficient distance λ to gain enough energy to produce the damage when the scattering event takes place. The resulting TDDB lifetime as a function of the applied electrical field E can then be written as:

$$MTTF = \frac{A'}{E} \cdot \exp(-\gamma'(T)\sqrt{E} + \frac{\alpha}{E}) \qquad (2)$$

where A' is a constant, $\gamma'(T)$ again a field acceleration parameter, and α a threshold parameter. This threshold parameter has its origin in the fact that the electron path lengths to a collision within the dielectrics most likely follow an exponential distribution. This assumption results in a significant increase of the lifetime when accelerated data are extrapolated to actual operational conditions. However, since the field acceleration during testing is usually quite large, it is difficult to determine the threshold parameter. Long-term, package level testing as opposed to fast, wafer-level testing is required to distinguish the models from each other. Figure 6 shows a comparison of various possible models [1]. As can be seen, regular wafer-level testing is not able to determine the true nature of the underlying TDDB model. Other techniques, namely *in-situ* techniques that allow to study the kinetics of the degradation process, are required to gain information on the true physical failure mechanisms.

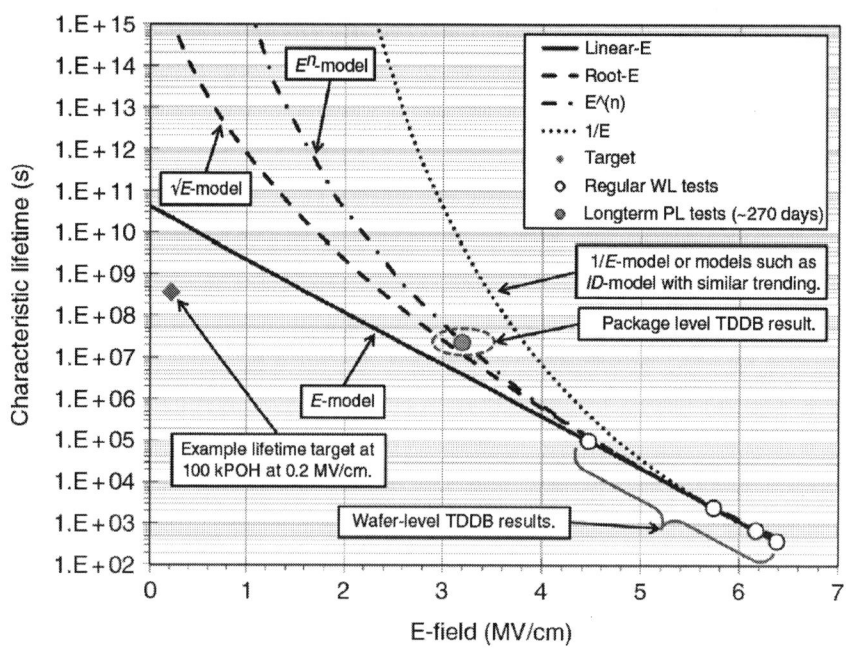

FIGURE 6. Comparison of several lifetime extrapolation models. In practice, wafer-level TDDB testing cannot distinguish between the different model approaches [1].

EFFECTS OF EXTERNAL STRESSES

In addition to the above described issues with the model development for TDDB failures, nearly no information is available on the effects of additional mechanical stresses caused by advanced packaging including 3D IC Integration or Chip-Package-Interaction (CPI). Typically, wafer processing can cause several 100 MPa of mechanical stress in the Si substrates, which in turn can alter the device and the interconnect performance and reliability [13]. Due to the 3D IC integration, this stress level can be preserved during operation and must be considered in performance and reliability evaluations for highly advanced, stacked integrated circuits (ICs). Figure 7 shows an example of a 3D-stacked IC system where two dies (Tier1 and Tier2) are stacked vertically and connected by Through-Silicon-Vias (TSVs), which typically consist of Cu [14]. Due to the mismatch in the Coefficients of Thermal Expansion (CTEs) between the Si substrate and the Cu TSV, significant stress levels occur in the 3D stack [15]. These stresses are highly dependent on the specific details process, materials used and geometries. The right of Fig. 7 shows a close-up of a local scenario where devices and Cu/low-k interconnect structures are located close to a TSV, with an interference of the stress field and the local wiring. These stress fields need to be taken into account in extrapolations to use conditions. The stress fields in the graph in Fig. 7 assume an oxide liner thickness of 0.5 µm at the periphery of the circular TSV, and are represented as a function of the diameter of the TSV. As expected, a larger TSV diameter leads to higher stress levels and a larger spatial extension. For an optimized use of the available chip area, designers need to know exactly how close a device and the respective interconnect structure can be placed towards a TSV. Both simulation and experimental studies are needed to provide this information to the design community.

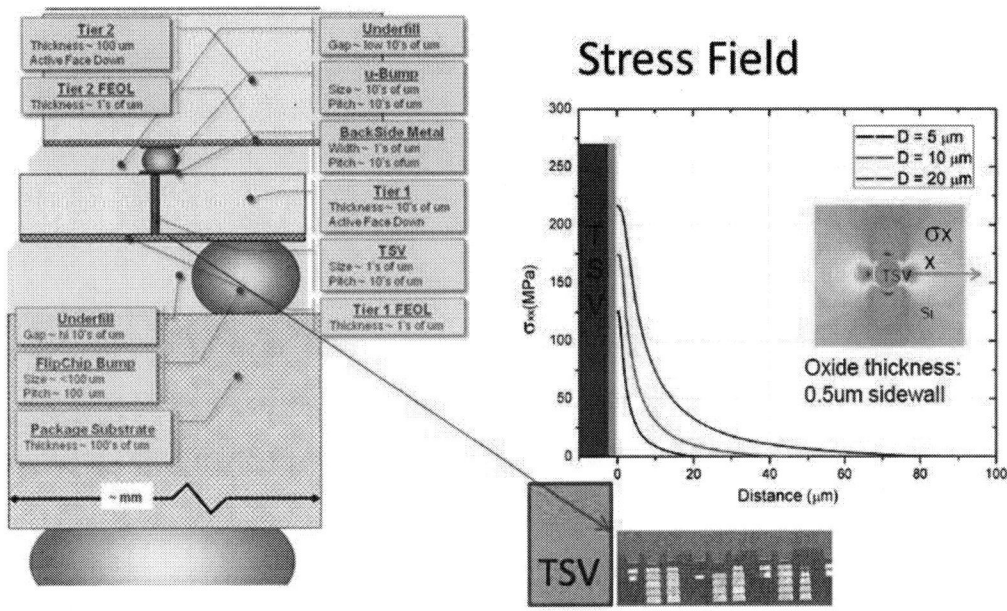

FIGURE 7. An example of 3D-stacked ICs on the left [14]. Tier1 and Tier2 dies have active devices facing down, connected by TSVs. The resulting stress field as shown on the right [15] can interact with the device and interconnect structures and needs to be taken into account in extrapolations of reliability test data to use conditions.

Considering the interference of mechanical stress fields with TDDB failure mechanisms, two possible scenarios can be envisioned. Firstly, highly localized stress effects could alter the physical mechanism. Fig. 8(a) shows a close-up of the top metal area where the intra-level dielectrics, the interconnect and the dielectric cap layer meet. The "cusp"-like barrier details can be seen. In this study, a buried cap layer (BCL), using a mechanically and electrically stronger dielectric film, was applied to enhance the TDDB performance [16]. A local change of the stress state at that position can change the way Cu ions are injected into the critical cap/dielectrics interface, assuming that this is the main failure mechanism and diffusion path [10]. Fig. 8(b) shows electrical field simulations for this metal-to-metal test structure and also points to the importance of the "cusp"-like area at the barrier interface [16]. The highest local electrical field is found just at that critical junction. Furthermore, if diffusion along the dielectrics interface is involved, the stresses along that interface will also influence the kinetics. Here, it is expected that the diffusional process will be altered according to the following equation [17]:

$$MTTF = A \cdot \exp(-\gamma(T) \cdot \sqrt{E}) \cdot \exp\left(\frac{E_A - \sigma\Omega}{k_B \cdot T}\right) \tag{3}$$

which is a modified version of Eq.(1), substituting the simple activation energy by the sum of E_a and a term that describes the effect of the additional stress field. σ denotes the mechanical stress in the respective direction and Ω the atomic volume.

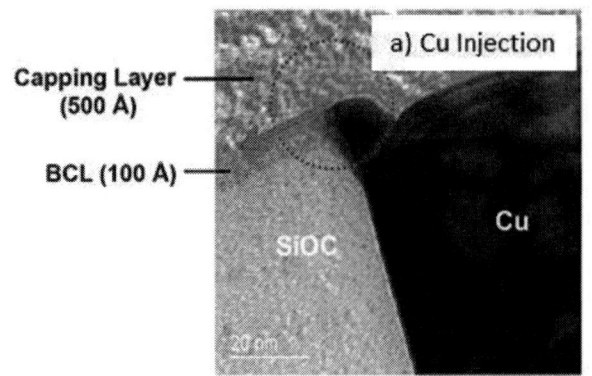 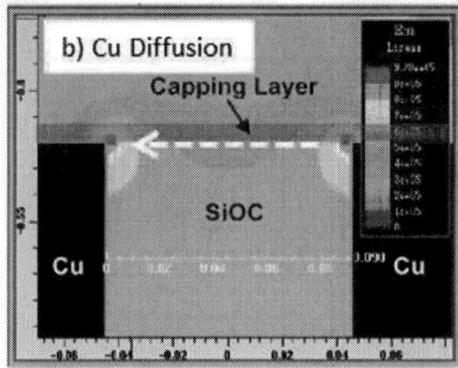

FIGURE 8. Cu injection (a) and diffusion (b) along the critical cap/dielectrics interface. Both mechanisms will be influenced by externally applied mechanical stress fields as shown in Fig. 7. In case (a), the stress effects are expected to be highly localized; in case (b), the entire cap/dielectrics interface is affected, and the diffusion kinetics is altered.

Table 1 estimates the effect of the added mechanical stress on the TDDB lifetime at a typical operational temperature of 125°C. Even at a relatively low stress level of about 200 MPa, the *MTTF* is expected to decrease by about 35%, which is a significant amount keeping in mind that target lifetimes are sometimes barely met during qualification efforts, especially considering large test structures (several meters of facing line length and millions of vias connected in chains). For a detailed characterization of these effects, new approaches have to be developed. Localized test structures should be used for investigations of the basic physical failure mechanisms. These localized structures, e.g. two facing metal ends or vias, will enable the characterization of TDDB kinetics, especially the behavior of the dielectrics and the movement of Cu ions. This should enable a clear definition of the underlying physics-based models and the correct extrapolation methods. Certainly, *in-situ* testing is one, maybe the only way to identify the true mechanism and to validate models. In addition, tests need to be performed applying external mechanical stress, both on localized regions as well as on extensive test structures to characterize both local and global effects.

TABLE 1. Effect of mechanically induced stress (3D IC and/or CPI related) on the TDDB lifetime. A stress level of 200 MPa is highlighted.

Stress (GPa)	Diffusivity Ratio at 125°C	*MTTF* Decrease
0.0	1.00	1.00
0.1	1.24	0.80
0.2	1.55	0.65
0.3	1.93	0.52
0.4	2.40	0.42
0.5	2.98	0.34
0.6	3.71	0.27
0.7	4.62	0.22
0.8	5.75	0.17
0.9	7.15	0.14
1.0	8.90	0.11

CONCLUSIONS

In addition to uncertainties in the actual physical mechanisms leading to TDDB failures and the resulting questions about the correct extrapolation models, very little is known about the impact of superimposed mechanical stress fields due to 3D IC integration. It is expected that additional stress fields can have a significant effect on the TDDB product lifetimes, similarly to the impact on transistor parameters. Therefore, novel approaches for the validation of models and for the identification of the true failure mechanism - including the effect of external stresses - need to be developed.

ACKNOWLEDGMENTS

The authors would like to thank Oliver Aubel (GLOBALFOUNDRIES Dresden Module One LLC & Co. KG, Dresden, Germany) and Ennis Ogawa (Broadcom Corporation, Irvine, CA, USA) for fruitful discussions and support. In addition, financial support from the State of Saxony is gratefully acknowledged (Project Number 66047/1135).

REFERENCES

1. E. Ogawa and O. Aubel, "Electrical Breakdown in Advanced Interconnect Dielectrics" in *Advanced Interconnects for ULSI Technology*, First Edition, edited by M.R. Baklanov, P. Ho and E. Zschech, John Wiley & Sons, Ltd. 2012, pp. 369-434.
2. J.D. Meindl, J.A. Davis, P. Zarkesh-Ha, C.S. Patel, K.P. Martin, and P.A. Kohl, IBM J. Res. Develop., 46 (2/3), pp. 245–263 (2002).

3. S. List, M. Bamal, M. Stucchi, and K. Maex, *Microelectron. Eng.* 83 (11/12), 2200–2207 (2006).

4. D. Edelstein, J. Heidenreich, R. Goldblatt, W. Cote, C. Uzoh, N. Lustig, P. Roper, T. McDevitt, W. Motsiff, A. Simon, J. Dukovic, R. Wachnik, H. Rathore, R. Schulz, L. Su, S. Luce, and J. Slattery, Tech. Dig., IEEE International Electron Devices Meeting, pp. 773–776 (1997).

5. S. Venkatesan, A.V. Gelatos, V. Misra, B. Smith, R. Islam R, J. Cope, B. Wilson, D. Tuttle, R. Cardwell, S. Anderson, M. Angyal, R. Bajaj, C. Capasso, P. Crabtree, S. Das, J. Farkas, S. Filipiak, B. Fiordalice, M. Freeman, P.V. Gilbert, M. Herrick, A. Jain, H. Kawasaki, C. King, J. Klein, T. Lii, K. Reid, T. Saaranen, C. Simpson, T. Sparks, P. Tsui, R. Venkatraman, D. Watts, E.J. Weitzman, R. Woodruff, I. Yang, N. Bhat, G. Hamilton, Y. Yu, Tech Dig., IEEE International Electron Devices Meeting, pp. 769–772 (1997).

6. K. Maex, M.R. Baklanov, D. Shamiryan, F. Iacopi, S.H. Brongersma, and Z.S. Yanovitskaya, *J. Appl. Phys.* **93** (11), 8793–8841 (2003).

7. W. Volksen, R.D. Miller, and G. Dubois, *Chem. Rev.* **110** (1), 56–110 (2010).

8. E.T. Ogawa, J. Kim, G.S. Haase, H.C. Mogul, and J.W. McPherson, Proc. IEEE International Reliability Physics Symposium, pp. 166–172 (2003).

9. O. Aubel, IEEE International Reliability Physics Symposium Tutorial (2011).

10. F. Chen, O. Bravo, K. Chanda, P. McLaughlin, T. Sullivan, J. Gill, J. Lloyd, R. Kontra, and J. Aitken, Proc. IEEE International Reliability Physics Symposium, pp. 46–53 (2006).

11. J.R. Lloyd, E. Liniger, and T.M. Shaw, J. Appl. Phys. **98**, 084109 (2005).

12. J.R. Lloyd, Materials Research Symposium Tutorial, Spring Meeting (2006).

13. V. Sukharev and E. Zschech, Int. Workshop on "Stress Management for 3D ICs Using Through Silicon Vias", Albany/NY 2010, AIP Conf. Proc. 1378, pp. 21–49 (2011).

14. R. Radojcic, M. Nowak, and M. Nakamoto, Int. Workshop on "Stress Management for 3D ICs Using Through Silicon Vias", Albany/NY 2010, AIP Conf. Proc. 1378, pp. 5–20 (2011).

15. P.S. Ho, S.-K. Ryu, T. Jiang, Q. Zhao, K.H. Lu, J. Im, and R. Huang, 12th International Workshop on Stress-Induced Phenomena in Microelectronics, May 2012, Kyoto, Japan (to be published).

16. K.Y. Yiang, T.S. Mok, W.J Yoo, and A. Krishnamoorthy, Proc. IEEE International Reliability Physics Symposium, pp. 333-337 (2004).

17. P. Shewmon, *Diffusion in Solids*, 1989.

Electromigration Void Nucleation and Growth Analysis Using Large-Scale Early Failure Statistics

M. Hauschildt[1], M. Gall[2], C. Hennesthal[1], G. Talut[1], O. Aubel[1], K.B. Yeap[2], E. Zschech[2]

1) GLOBALFOUNDRIES Dresden Module One LLC & Co. KG, Wilschdorfer Landstr. 101, D-01109 Dresden, Germany
2) Fraunhofer Institute for Nondestructive Testing (IZFP-D), Maria-Reiche-Str.2, D-01109 Dresden, Germany

Abstract. Electromigration early failure void nucleation and growth phenomena were studied using large-scale, statistical analysis methods. A total of about 470,000 interconnects were tested over a wide current density and temperature range (j = 3.4 to 41.2 mA/μm^2, T = 200 to 350°C) to analyze the behavior of the current density exponent n as a function of temperature. The results for the critical V1M1 downstream interface indicate a reduction from n = 1.55±0.10 to n = 1.15±0.15 when lowering the temperature towards 200°C for Cu-based interconnects. This suggests that the electromigration downstream early failure mechanism is shifting from a mix of nucleation-controlled (n = 2) and growth-controlled (n = 1) to a fully growth-controlled mode, assisted by the increased thermal stress at lower temperatures (especially at use conditions). For Cu(Mn)-based interconnects, a drop from n = 2.00±0.07 to n = 1.60±0.17 was found, indicating additional effects of a superimposed incubation time. Implications for extrapolations of accelerated test data to use conditions are discussed. Furthermore, the scaling behavior of the early failure population at the NSD = -3 level ($F \sim 0.1\%$) was analyzed, spanning 90, 65, 45, 40 and 28 nm technology nodes.

Keywords: Electromigration, Cu Interconnects, Early Failure, Void Nucleation, Void Growth, Current Density Exponent, Large-Scale Statistics

INTRODUCTION

Advanced interconnects continue to show electromigration (EM) concerns due to ever decreasing geometries with current carrying demands remaining high. EM failures are typically observed at the cathode end of interconnect lines. At this point, where a via connects two metal levels, a barrier layer creates a flux divergence site for Cu. The complexity of typical EM analysis was significantly increased after both up- and downstream tests started to show bimodality in the lifetime distributions [1, 2]. Early failures were either attributed to voids in the vias (upstream test) or slit voids under the via (downstream test), whereas the later failures resulted from larger faceted voids spanning most of the metal line. Furthermore, EM tests are performed at higher temperatures and current densities compared to operating conditions to obtain lifetime distributions in a reasonable time. As a consequence, extrapolations from EM stressing conditions to operating conditions according to the following equation are necessary

$$TTF_{oper} = MTTF_{stress} \left(j_{stress} \Big/ j_{oper} \right)^n \exp\left[\frac{E_a}{k} \left(\frac{1}{T_{oper}} - \frac{1}{T_{stress}} \right) + NSD * \sigma \right] \quad (1)$$

where TTF is the time to failure, $MTTF$ the median lifetime, j the current density, n the current density exponent, E_a the activation energy, k Boltzmann's constant, T the temperature, NSD the number of standard deviations, and σ the lognormal standard deviation. The subscripts 'stress' and 'oper' refer to EM testing and actual operating conditions, respectively. Besides the median lifetimes and lognormal standard deviation values of experimentally acquired failure time distributions, essential parameters to calculate from testing temperature and current density to operating values are the activation energy related to the dominating diffusion mechanism and the current density exponent. Both parameters have been extensively studied for the main portion of the EM lifetime distribution. For pure Cu, E_a is usually found to be ~0.9eV for lines capped with SiCN and slightly higher (~1.0eV) for interconnects capped with SiN [3]. The dominant diffusion path in these structures is the top interface of the Cu lines, whereas diffusion along Cu/barrier layer interfaces or Cu grain boundaries appeared to be negligible [4, 5, 6]. In general, a correlation between the work of adhesion for the Cu line/cap layer interface and E_a was observed with E_a increasing for larger adhesion values [7]. In contrast, the reported n-values for the main portion of the lifetime distribution vary significantly between 1 and 2 [4, 8]. For extrapolation calculations, conservative values of $Ea \sim 0.9eV$ and $n \sim 1$ are usually being used. Unfortunately, this practice leads to quite some uncertainty in the extrapolation results and possibly puts unnecessary constraints on designers as well as the process.

Furthermore, the already mentioned bimodality causes additional concern whether the generally used values are appropriate. It was already shown that the activation energy for the early and late failure modes can be significantly different with the downflow early mode showing a slight, but significant reduction to 0.83 ± 0.01 eV [9]. Also, another study seemed to indicate that the n values for early and late modes can differ significantly [10]. The early failure mode yielded values closer to $n = 2$, whereas the late mode was characterized by $n \sim 1$. Hence, the overall objective of this study is the evaluation of activation energy and current density exponent for the early mode over a broad temperature and current density range, i.e. this study intends to answer the question whether E_a and n are constant between operating and use conditions. The focus of this paper is on the current density exponent. In addition to experimental results using large scale EM tests (Wheatstone bridge methodology, WSB), modeling was performed assuming that the current density dependence is based on a mix of nucleation-controlled ($n = 2$) and growth-controlled ($n = 1$) processes.

EXPERIMENTAL PROCEDURE

Standard EM testing is performed on a Cu line connected by vias on both ends of the structure to a lower or upper metal level. The current flows from a wide metal line

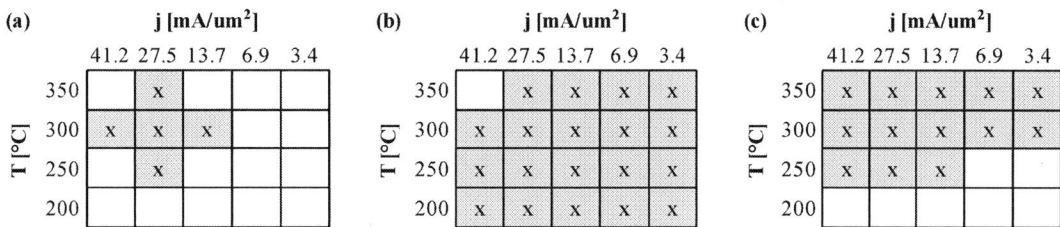

FIGURE 1. EM test matrices to evaluate Ea and n: (a) generally used for regular single-link EM structures, and for WSB devices (b) with pure Cu interconnects and (c) with Cu(Mn) lines

through the via into the test line. If the test line is on a lower metal level compared to the current supply line, the test is called 'downstream', whereas it is called 'upstream' for test lines on a higher metal level. Only one via/line interface is being tested per sample. A typical matrix of EM tests generally conducted to evaluate E_a and n is shown in Fig. 1(a). Due to long lifetimes of single-link interconnects, additional tests to increase the current density and temperature range are rarely feasible. Thus, this study is based on the WSB methodology, where each sample contains a large number of parallel and serial interconnects assembled in a Wheatstone bridge arrangement. Due to the extremely high resolution of this wiring, the smallest changes in voltage and resistance can be detected as an imbalance in the bridge. These small changes correspond to the first voids, which form in any of the via/interconnect elements within the entire bridge. Besides significantly reducing testing time and thus enabling the evaluation of more testing conditions, the use of WSB structures provides data for the early failure mode, which defines the chip lifetime. Details describing this methodology were published previously [9, 11]. The EM test matrices using WSB devices are shown in Fig. 1(b) using pure Cu interconnects and Fig. 1(c) using Cu(Mn) lines. Due to differences in lifetime, some adjustments of the testing conditions were needed, but between three and five current conditions were evaluated at three or four temperatures resulting in the evaluation of very wide temperature and current density ranges (200-350°C and 3.4-41.2 mA/μm^2). For each cell about 12 WSB samples with 1080 vias each were tested, representing a total of ~ 12500 vias per cell. The total sum was ~237600 tested Cu vias and ~233000 Cu(Mn) vias. In this study, V1M1 (downstream) samples from the 40 nm technology generation were used.

Due to the close proximity of lines under test in WSB structures, current-induced heating needs to be considered. Joule heat was evaluated by standard methods, i.e. the resistance was measured as a function of current at various temperatures. Figure 2 shows the difference in temperature as a function of current density at different stress temperatures. Joule heat of up to 12°C for the largest current density and temperature can be observed. Obviously this effect cannot be neglected and all lifetime data is Joule heat corrected. It needs to be noted that this methodology results in an average Joule heat value. Detailed simulations are needed to evaluate possibly existing local hot spots.

Figure 3 represents a plot of lifetime distributions for downstream WSB devices, 20-link structures and the corresponding single link samples tested at the same temperature and current density. The line through the single link data using a monomodal lognormal fit is for illustrative purposes only. In this chart, the multi-link data has been deconvoluted, i.e. the non-failed links in the EM failure distribution have been included in the data analysis as censored values to enable a direct comparison to the single-link data. Details of this calculation were reported earlier [9, 11]. The lifetime

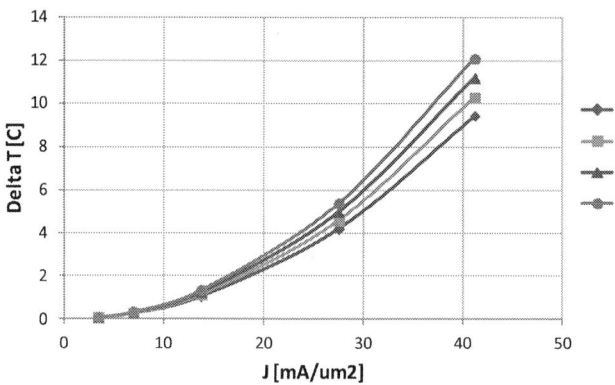

FIGURE 2. The deviation from oven temperature due to Joule heat as a function of current density at different stress temperatures

distributions of differently sized test structures line up well. An early fail tail, not easily detectable with single links, is clearly visible. Thus, the application of WSB devices enables "deep censoring" in the very early failure regime. All further data presented in this paper is based on WSB devices, and hence represents the characteristics of the early failure mode, which is the most critical for the overall chip lifetime.

RESULTS

Typical failure distributions for 5 different current densities are shown in Fig. 4(a) for Cu and Fig. 4(b) for Cu(Mn), respectively. Note that this data is not deconvoluted and one data point represents the first fail in 1080 interconnects. The direct comparison to single-link results is not necessary here. The test temperature was 300 °C and the current densities ranged from 3.4 to 41 mA/μm^2. All lifetime distributions show reasonable *MTTF* and sigma values. Compared to pure Cu, the lifetimes of Cu(Mn) interconnects are significantly longer as already reported in various studies [12-15]. Segregation of Mn towards the most critical locations, such as the interface to the upper dielectric layer, as well as an enhancement effect of imperfections at the liner, are generally thought to enable the significantly better EM performance. The

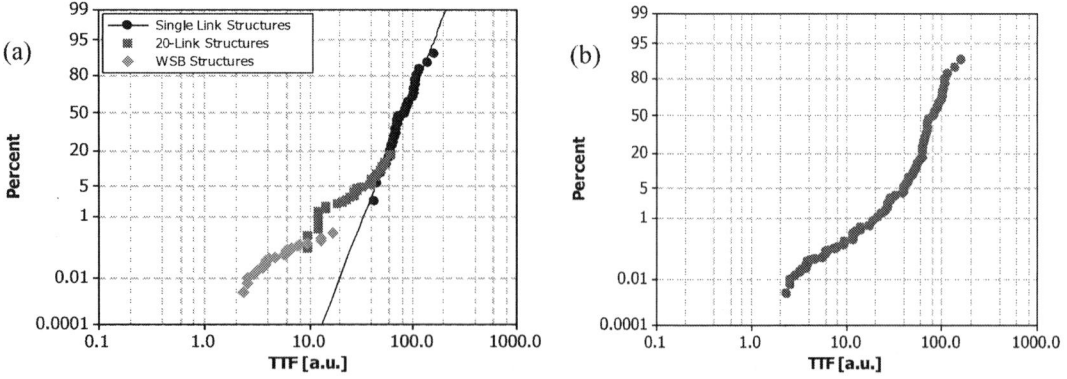

FIGURE 3. Lifetime distributions for downstream Wheatstone Bridge devices, 20-link structures and the corresponding single link samples: (a) as separate distributions and (b) all data combined

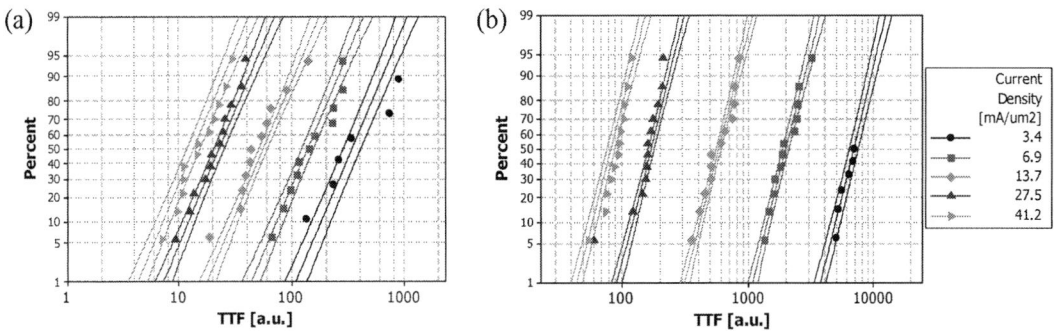

FIGURE 4. Typical failure distributions for 5 different current densities (a) for Cu and (b) for Cu(Mn)

lifetime distributions of the other test cells are equally good, thus acceptable error bounds were achieved for the *n*-value extraction. Current exponent values as a function of temperature are listed in Table 1 for both Cu and Cu(Mn) interconnects. A change in the *n*-value as a function of temperature is clearly visible, with *n* dropping significantly with a decrease in temperature. Interestingly, the *n*-values for Cu(Mn) are approximately 0.45 higher compared to Cu. Considering these results, extrapolations using *n* > 1 have to be treated with caution, but using *n* ~ 1 may be too conservative imposing too many restrictions on designers and the manufacturing process.

MODELING AND ANALYSIS

Current density values between *n* = 2 and *n* = 1 can possibly be explained when dividing the EM process into two parts: nucleation and growth [16]. The *MTTF* value can then be described as the sum of nucleation time t_n and the time of void growth, t_{growth}. The nucleation term can be understood as the time it takes to build up a certain "threshold stress" in tension for voids to form, σ_n:

$$t_n = \gamma kT \exp(\frac{E_a}{kT})(\frac{\sigma_n}{j})^2 \tag{2}$$

where γ is a constant. This stress can be interpreted in various ways, such as an effect of vacancy supersaturation or local delamination stress, e.g. right under the via. Basically, Cu ions diffusing due to EM leave vacancies behind. The resulting space is being filled by the surrounding Cu atoms to reduce the overall energy of the system. Once a certain number of Cu ions have moved, it is energetically more favorable to form a void instead of further stretching the bonds of the surrounding metal. As can be seen from Eq. 2, the stress term leads to a current density exponent of *n* = 2. The EM void growth term follows the usually assumed drift behavior

TABLE 1. Current exponent values as a function of temperature for Cu and Cu(Mn) interconnects

Temperature [°C]	n [Cu lines]	n [Cu(Mn) lines]
350	1.55 ± 0.10	2.00 ± 0.07
300	1.30 ± 0.10	1.78 ± 0.05
250	1.21 ± 0.16	1.60 ± 0.17
200	1.15 ± 0.15	

$v_d = \mu Z^* e\rho j$, where v_d is the drift velocity, μ the atomic mobility, Z^* the effective charge number, e the electric charge and ρ the metal resistivity. As a result, the void growth term follows a current density exponent of $n = 1$:

$$t_{growth} = \frac{C}{\mu Z^* e\rho j} = (\frac{A'kT}{j})\exp(\frac{E_a}{kT}) \qquad (3)$$

where C and A' are constants. In addition to these basic terms, the thermal stress σ_{th} in the interconnect structure needs to be considered:

$$\sigma_{th}(T) = E\Delta\alpha(T_0 - T) \qquad (4)$$

where E is the appropriate effective modulus, $\Delta\alpha$ the difference in the thermal expansion coefficients between Cu and the surrounding encapsulation, and T_0 the stress-free temperature for Cu [16]. Generally it is assumed that the temperature at which the Cu interconnect attains zero stress is determined by the highest temperature applied during processing. One study of thermal stresses in Cu interconnects embedded in SiCOH shows that the zero stress state occurs approximately at 400°C [17], which is in good agreement with the above assumption. Thus, at high temperatures, the thermal stress is small and does not interfere with the EM mechanism. However, with decreasing temperatures, the thermal stress increases considerably in the tensile direction. It can interact with the EM mechanism by reducing the time it takes to reach the "threshold stress" for void formation. Equation 5 describes the resulting reduction in nucleation time:

$$t_n = \gamma kT\exp(\frac{E_a}{kT})(\frac{\sigma_n - E\Delta\alpha(T_0 - T)}{j})^2 \qquad (5)$$

Thus, a more growth-controlled EM process is expected at lower temperatures. Combining Equations 3 and 5 results in the following expression for the total EM lifetime:

$$MTTF = t_n + t_{growth} = \frac{B'(T)kT}{j^2}\exp(\frac{E_a}{kT}) + \frac{A'kT}{j}\exp(\frac{E_a}{kT}) \qquad (6)$$

where B'(T) describes the temperature-dependent effect of the thermal stress in the interconnects. In order to separate nucleation and growth processes in the experimental EM data, Eq. 6 can be further simplified to

$$MTTF = \frac{B(T)}{j^2} + \frac{A(T)}{j} \qquad (7)$$

B(T) and A(T) values can be determined from experimental data for all test temperatures. An example is shown in Fig. 5. Two different fits are included, one using a single, effective n and the other a separation into $n = 1$ and $n = 2$. Both fits are equally good. Arrhenius plots for A(T)/kT and B(T)/kT are shown in Fig. 6. The growth term A/kT seems to follow a straight Arrhenius behavior with an activation energy of ~ 1.0 eV, very close to the expected value for Cu diffusion along the SiCN interface [3]. The nucleation term B/kT shows a clear deviation from a straight Arrhenius behavior with the deviation increasing for high temperatures. Since both A/kT and B/kT include an Arrhenius term $exp(E_a/kT)$, the ratio of B/A represents the effect of the thermal stress, or any other additional effects besides the regular Arrhenius-type behavior. Equation 8 describes the proposed stress term:

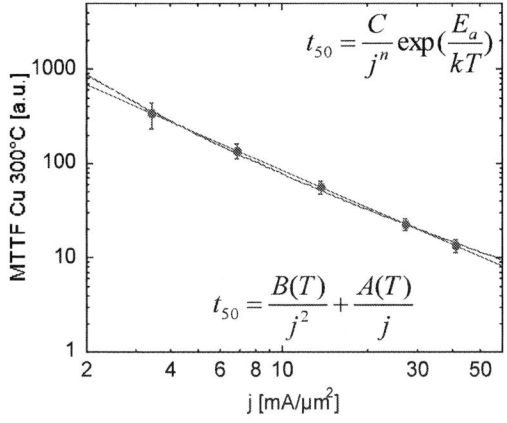

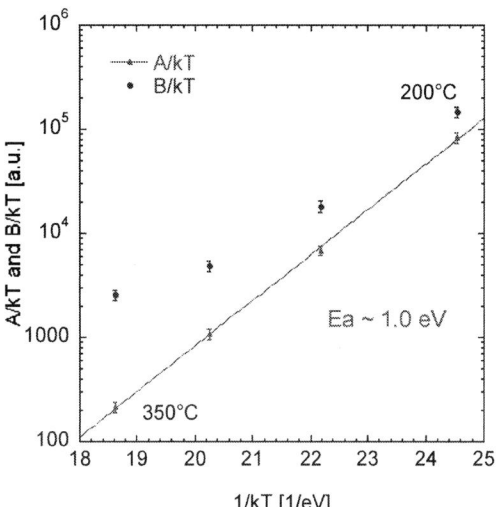

FIGURE 5. *B(T)* and *A(T)* values determined from experimental data at 300°C. Two different fits are included, one using a single, effective *n* and the other a separation into *n=1* and *n=2*

FIGURE 6. Arrhenius plots for *A(T)/kT* and *B(T)/kT*

$$\frac{B}{A} \propto (\sigma_n - E\Delta\alpha(T_0 - T))^2 \tag{8}$$

which can be further simplified to

$$\frac{B}{A} \propto (T - T_{th=n})^2 \tag{9}$$

where $T_{th=n}$ represents the temperature at which thermal and nucleation stresses coincide. Figure 7 shows a fit to the experimental data using Eq. 9. A good fit is obtained for the temperature range of 200°C to 300°C, but the strong increase of B/A measured at 350°C is not well-reproduced. From Figure 7 it can be determined that the thermal stress and the nucleation stress coincide at a $T_{th=n}$ of approximately 50°C (323K). Several studies have evaluated the stress state in Cu interconnects as a function of temperature. Laboratory-based X-ray measurements show a hydrostatic stress state of about 360 MPa at room temperature in Cu interconnects of 200nm width embedded in SiCOH dielectric [17]. Synchrotron X-ray studies using Cu line structures with SiCOH dielectric of widths between 50 and 500nm resulted in a hydrostatic stress state of approximately 375 MPa at room temperature for 66nm wide lines [18]. Thus, at a $T_{th=n}$ value of 50°C, the stress state in the Cu interconnect is expected to be approximately 340 MPa hydrostatic. This value appears reasonable for the critical void nucleation stress.

Figure 8 shows the model prediction of *n* as a function of temperature. As in Figure 7, only the temperature range up to 300°C is well-described. Hence, the assumption that the void nucleation term has a square-dependence on the threshold stress seems to be rather appropriate for a temperature range up to 300°C. However, additional mechanisms may play a role between 300 and 350°C, such as stress relaxation effects that counteract the build-up of the threshold stress. Further refinement for Cu and modeling efforts for Cu(Mn)-based interconnects are ongoing, addressing incubation

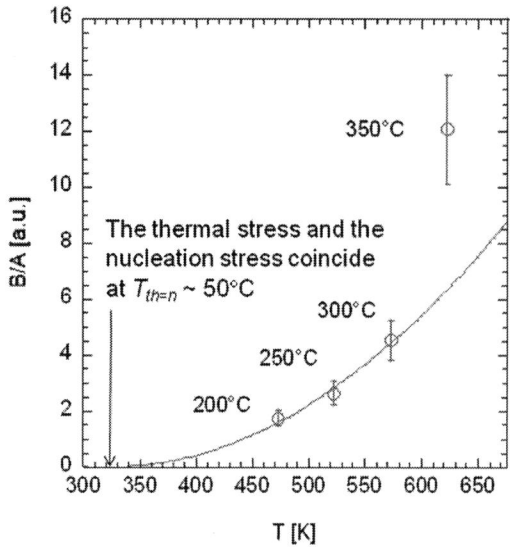

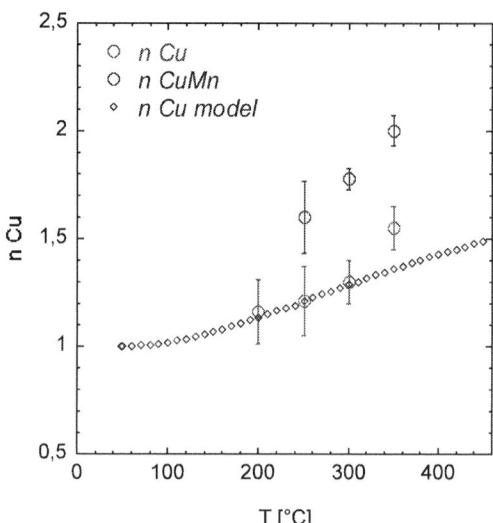

FIGURE 7. Experimental data for B/A as a function of temperature and corresponding fit using Eq. 9. B/A represents the effect of the thermal stress, or any other additional effects besides the regular Arrhenius-type behavior

FIGURE 8. Current exponent values as a function of temperature for Cu and Cu(Mn) interconnects including the model prediction of n for Cu lines

time effects shifting n towards 2. Clearly, after successful modeling efforts for the temperature-dependent behavior of the current density exponent, an effective value has to be determined. This effective value will depend both on the temperature range used to span test and operating conditions, as well as the current density range. Furthermore, initial results of this study indicate that the activation energy is not a constant when analyzed as a function of current density, as predicted earlier [16]. Again, an effective activation energy needs to be determined, adding quite some complexity to the extrapolation methods. None of the typically used extrapolation parameters for EM performance are constant, but interdependencies exist.

TECHNOLOGY SCALING

The major part of this paper deals with the current density exponent for the early failure mode of one technology node in order to obtain a detailed understanding on the proper value which needs to be used to extrapolate from EM test conditions to operating conditions. This chapter briefly examines the evolution of EM lifetimes of the early failure mode through various technology generations. Figure 9 shows EM lifetimes at about 0.1% cumulative failure (NSD = -3) as a function of the product of the line height and via size for various technologies. These lifetime values at low percentage levels were directly obtained from measurement through the use of multilink structures. With various products requiring only a cumulative failure target of 0.1%, the need for extrapolations using Eq. 1 are reduced to scaling from EM test current and temperature. Besides enabling "deep censoring" in the very early failure regime, the application of WSB devices is extremely efficient also in terms of testing time, i.e. it is up to 10x faster.

As expected, the EM lifetime of Cu drops for every technology generation. Depending on the current carrying needs, the introduction of EM enhancing processes was clearly needed at some point below 65nm technologies. Cu(Mn) significantly enhances the EM lifetime for 40nm and 28nm technologies to levels exceeding the 65nm value.

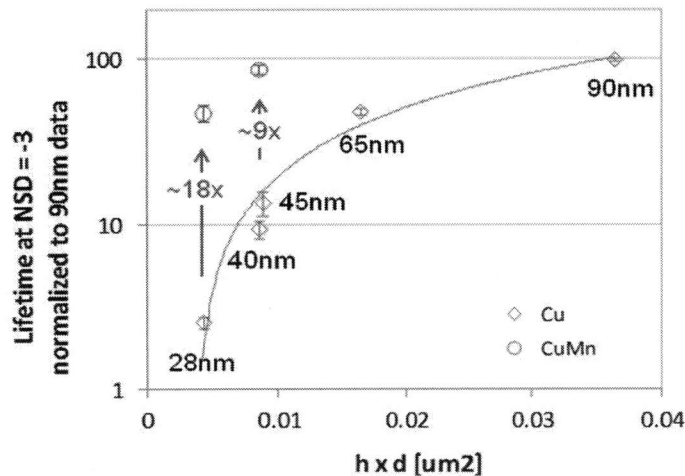

FIGURE 9. Lifetime at 0.1% (NSD = -3) as a function of line height *h* x via size *d* for various technologies

For the 28nm node, Cu(Mn) lifetime improvement over pure Cu was found to be about 18x for the early EM failure mode. While Cu(Mn) appears to be a reasonable candidate for the 20nm technology generation, further process enhancements, such as metal caps, might be necessary for 14nm nodes and beyond.

CONCLUSION

A clear understanding of the current density exponent *n* is necessary for correct extrapolations from EM testing to operating conditions. Using large-scale statistical analysis methods, a significant decrease of *n* with lowered temperature was observed for Cu as well as Cu(Mn) interconnects. The EM downstream early failure mechanism can be separated into void nucleation and growth stages, leading to two separate current density exponents. The observed decrease of *n* with reduced temperature suggests a shift from a mix of nucleation-controlled ($n = 2$) and growth-controlled ($n = 1$) to a fully growth-controlled mode, assisted by the increased thermal stress at lower temperatures. Modelling efforts for Cu were successful between room temperature and 300°C, describing the temperature behavior of *n* very well. Additional mechanisms, such as stress relaxation effects, may play a role at higher temperatures. Furthermore, it is expected that effects of a superimposed incubation time will need to be considered for Cu(Mn)-based interconnects, which showed the same trends for *n*, but generally higher values. Considering these results, the *n* value used for lifetime extrapolations needs to be carefully selected. Using $n \sim 1$ is a conservative approach. In the case that this value imposes too many restrictions on designers and process, extrapolations using an effective $n > 1$ can be explored with caution.

REFERENCES

1. J. Gill, *et al.*, IEEE Int. Reliab. Phys. Symp. Proc. (2002), p. 298
2. B. Li, *et al.*, IEEE Int. Reliab. Phys. Symp. Proc. (2006), p. 115
3. M. Gall, *et al.*, Mater. Res. Soc. Symp. Proc. 914 (2006), p. 305
4. C.-K. Hu, *et al.*, IEEE Int. Interconnect Tech. Conf. Proc. (1999), p. 267
5. C.S. Hau-Riege, *et al.*, Appl. Phys. Lett. 78 (22) (2001), p. 3451
6. E.T. Ogawa, *et al.*, IEEE Trans. on Rel. 51 (4) (2002), p. 403
7. J.R. Lloyd, *et al.*, IEEE IRW Final Report (2002), p. 32
8. M. Hauschildt, *et al.*, AIP Conf. Proc. of Stress Induced Phenomena in Metallization: 9th Int. Workshop, 945 (2007), p. 66
9. M. Hauschildt, *et al.*, J. Appl. Phys. 108, 013523-1-10 (2010)
10. R. Filippi, *et al.*, IEEE Int. Reliab. Phys. Symp. Proc. (2009), p. 444
11. M. Gall, *et al.*, J. Appl. Phys. 90 (2) (2001), p. 732
12. J. Koike, *et al.*, IEEE Int. Interconnect Tech. Conf. Proc. (2006), p. 161
13. C. Christiansen, *et al.*, IEEE Int. Reliab. Phys. Symp. Proc. (2011), p. 312
14. C. Christiansen, *et al.*, IEEE Int. Reliab. Phys. Symp. Proc. (2012), 5E.1.1
15. C.-K. Hu, *et al.*, J. Appl. Phys. 111, 093722 (2012)
16. J.R. Lloyd, Microelectron. Reliab. 47, 1468 (2007)
17. D.W. Gan, *et al.*, IEEE Int. Interconnect Tech. Conf Proc. (2002), p. 271
18. C. Wilson, *et al.*, J. Appl. Phys. 106, 053524 (2009)

Modeling of Microstructural Effects on Electromigration Failure

H. Ceric[*], R. L. de Orio[†], W. Zisser,[†] and S. Selberherr[†]

[*]*Christian Doppler Laboratory for Reliability Issues in Microelectronics at the Institute for Microelectronics, TU Wien*
[†]*Institute for Microelectronics, TU Wien, Gußhausstraße 27–29, A-1040 Wien, Austria*

Abstract. Current electromigration models used for simulation and analysis of interconnect reliability lack the appropriate description of metal microstructure and consequently have a very limited predictive capability. Therefore, the main objective of our work was obtaining more sophisticated electromigration tools. The problem is addressed through a combination of different levels of atomistic modeling and already available, continuum level macroscopic models. A novel method for an *ab initio* calculation of the effective valence for electromigration is presented and its application on the analysis of EM behavior is demonstrated. Additionally, a simple analytical model for the early electromigration lifetime is obtained. We have shown that its application provides a reasonable estimate for the early electromigration failures including the effect of microstructure. A simulation study is also applied on electromigration failure in tin solder bumps, where it contributed the understanding of the role of tin crystal anisotropy in the degradation mechanism of solder bumps.

Keywords: EM, interconnect, reliability, physical modeling, simulation
PACS: 66.30.Qa

INTRODUCTION

Electromigration (EM) experiments indicate that the interconnect lifetime decreases with every new interconnect generation. In particular, fast diffusivity paths cause a significant variation in the interconnect performance and EM degradation [1]. In order to produce more reliable interconnects, the fast diffusivity paths must be addressed when introducing new designs and materials. The EM lifetime depends on a variation of material properties at the microscopic and atomistic levels. Microscopic properties are grain boundaries and grains with their crystal orientation [2]. Atomistic properties are configurations of atoms at the grain boundaries, at the interfaces to the surrounding layers, and at the cross-section between grain boundaries and interfaces. Modern Technology Computer-Aided Design (TCAD) tools, in order to meet the challenges of contemporary interconnects, must cover two major areas: physically based continuum-level modeling and first-principle/atomistic-level modeling.

We present a computationally efficient *ab initio* method for calculation of the effective valence for EM and the atomistic EM force. The results of these *ab initio* calculation are applied for parameterization of a continuum-level model [3] and for simulation of the impact of the copper microstructure on the EM behavior. Additionally, an application of the kinetic Monte Carlo method in combination with the *ab initio* method for EM analysis is demonstrated.

99

Results of *ab initio* and atomistic calculations are also used for the derivation of a compact model for early EM failures in copper dual-damascene M1/via structures. The model is based on the combination of a complete void nucleation model together with a simple mechanism of slit void growth under the via. It is demonstrated that the early EM lifetime is well described by a simple analytical expression, from where its statistical distribution can be obtained. Moreover, it is shown that the simulation results provide a reasonable estimate for the EM lifetimes.

For the realization of modern three-dimensional (3D) integrated circuits new interconnect components such as through-silicon-vias (TSVs) and solder bumps, together with complex multi-level 3D interconnect structures are gaining importance. Solder bumps are important components for 3D integration, because they enable vertical stacking of wafers. Tin solder bumps often consist of several large grains, such that most solder bumps exhibit one or at most a few grain orientations [4]. Tin has a tetragonal crystal structure which exhibits highly anisotropic diffusional, mechanical, thermal, and electrical properties [5]. In the final part of this work, we study the different EM failure modes of solder bumps caused by a microstructural anisotropy of tin.

THEORETICAL BACKGROUND

Electronic Density Based Calculation of Effective Valence

Generally, the effective valence is a tensor field $(\bar{Z})$, which defines a linear relationship between the EM force $(\vec{F})$ and an external electric field $\vec{E}$.

$$\vec{F}(\vec{R}) = e\bar{Z}(\vec{R})\vec{E} \tag{1}$$

For the calculation of the effective valence several methods have been proposed, all of them being based on the computation of electron scattering states [6]. Density functional theory (DFT), in connection with the augmented plane wave (APW) method [7] or the Korringa-Kohn-Rostoker (KKR) method [8], has been established as the most powerful method for the determination of scattering states, however, it requires a demanding computational scheme. The cumbersome representation of scattering wave functions with many parameters is a heavy burden on stability and accuracy of subsequent numerical steps. In this work we introduce a more robust and efficient method to calculate the effective valence, which relies only on the electron density $\rho_d(\vec{k}, \vec{r})$. The basic idea is given in the following equations for the tensor components [9]:

$$Z_{i,j}(\vec{R}) = \frac{\Omega_c}{4\pi^3} \iiint d^3\vec{k}\, \delta(\mathscr{E}_F - \mathscr{E}_{\vec{k}}) \tau(\vec{k}) [\vec{v}(\vec{k}) \cdot \hat{x}_j] \cdot$$
$$\cdot \iiint d^3\vec{r}\, \rho_d(\vec{k}, \vec{r}) [\nabla_{\vec{R}} V(\vec{R} - \vec{r}) \cdot \hat{x}_i], \tag{2}$$

V is the Coulomb interaction potential between an electron and the migrating atom, $\tau(\vec{k})$ is the relaxation time due to scattering by phonons, $v(\vec{k})$ is the electron group velocity, $\vec{E}$ is the external electric field, $\mathscr{E}_F$ is the Fermi energy, $\mathscr{E}_{\vec{k}}$ is the energy which corresponds to a state vector $\vec{k}$, and Ω_c is the volume of a unit cell. The first integration is over the k-space and the second over the volume of the crystal. For the calculation of the electron density the DFT tool VASP [10] is used. An example of a VASP calculation is presented in Fig. 1. The electron density alone provides a qualitative explanation for the fact that the effective valence is higher in the bulk than in the grain boundaries. Higher electron

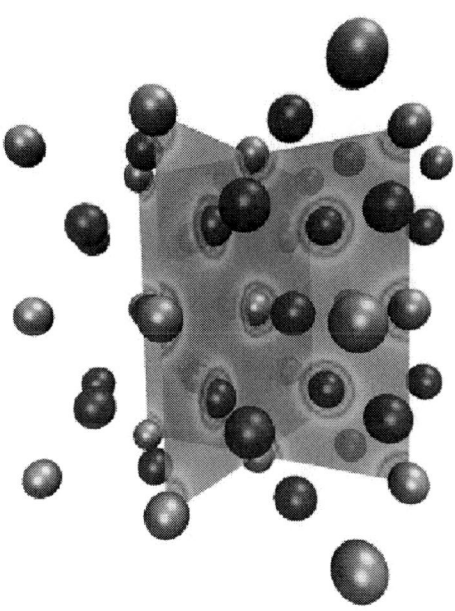

FIGURE 1. Portion of the bulk copper crystal. The electron density is represented in two orthogonal planes. It varies from higher values (circle regions around atoms) closer to the atomic nucleus to lower in the inter-atomic space.

densities lead to higher effective valences, as can be seen from (2). Similar analyses can be performed for atomic structures of different copper/insulator interfaces. For an accurate electron density calculation it is necessary to know the exact positions of the atoms in the structure.

Kinetic Monte Carlo Simulation of Electromigration

To utilize results of quantum mechanic calculations for kinetic Monte Carlo simulations an average driving force along the diffusion jump path must be calculated. In general, the microscopic force-field depends on the position of the defect along the diffusion jump-path. The average of the microscopic force over the j-th diffusion jump path between locations $\vec{r}_{j,1}$ and $\vec{r}_{j,2}$ [6] is

$$F_{m,j} = \frac{1}{|\vec{r}_{j,2} - \vec{r}_{j,1}|} \int_{\vec{r}_{j,1}}^{\vec{r}_{j,2}} \vec{F}(\vec{r}) \cdot d\vec{r}. \tag{3}$$

The change in diffusion barrier height $\Delta A_{\alpha j}$ is equal to the work done on the defect by the microscopic force as the defect is moved from initial to final sites over the entire jump path. The rates of defect jumps are calculated using the harmonic approximation to transition state theory (TST) [11]. In this approximation, the transition rate $\Gamma_{\alpha j}$ is given by

$$\Gamma_{\alpha j} = \nu_0 \exp\left(-\frac{E_m - \Delta A_{\alpha j}}{kT}\right). \tag{4}$$

E_m is the migration energy (barrier) defined as the difference in energy between the transition state and the initial state, and ν_0 is an attempt frequency [11]. For each defect site α, the residence time is calculated as [11]

$$\tau_\alpha = \frac{1}{\sum_{j=1}^{k_\alpha} \Gamma_{\alpha j}}. \tag{5}$$

k_α is the number of possible jump sites from site α. A single point defect is created at an arbitrary site, the clock is set to zero, and the defect is released to walk through the system. At each step the jump direction is decided by a random number according to the local jump probabilities

$$P_{\alpha j} = \tau_\alpha \Gamma_{\alpha j}. \tag{6}$$

The jump is implemented by updating the coordinates of the defect. By repeating the described random walk procedure for millions of defects, their concentration dependence on the effective valence tensor and the external field is calculated.

Compact Model for Lifetime Estimation

In order to calculate the mechanical stress in a three-dimensional copper dual-damascene interconnect structure, a complex physically based model including the EM equation, the electro-thermal equation, and the mechanical equations has to be solved [3].

Korhonen *et al.* [12] proposed a simple one-dimensional model, where the solution for the stress at the cathode of a semi-infinite line is given by

$$\sigma(t) = \frac{2eZ^*\rho j}{\Omega} \sqrt{\frac{D_a B\Omega}{\pi kT} t} = a\sqrt{t}. \tag{7}$$

D_a is the effective atomic diffusivity and B is the effective modulus, which depends on the metal and the surrounding materials.

Void formation occurs as soon as the mechanical stress reaches a critical magnitude at a site of weak adhesion, typically at the copper/capping layer interface [13], [14]. Thus, the void nucleation time is determined by the condition $\sigma(t_n) = \sigma_c$, which applied to (7) yields

$$t_n = \frac{\pi}{4} \frac{\Omega kT}{(eZ^*\rho j)^2 BD_a} \sigma_c^2 = \left(\frac{\sigma_c}{a}\right)^2, \tag{8}$$

where σ_c is the critical stress.

The solution given by (8) is a good approximation to the more complete solution obtained by solving a full physicall model [3],[15] numerically, as will be shown later. It should be pointed out that (8) is valid as long as the stress remains significantly smaller than the stress magnitude at the steady state condition, which holds true for the void formation phase.

Void Growth

For a copper dual-damascene M1/via structure with downstream electron flow, EM failure analyses [16] indicate that the early failures are caused by slit voids located under the via, as shown in Fig. 2. Since the void is very thin and does not grow through the line height, void growth can be described by a one-dimensional process, so that the void length is given by

$$l_{void} = v_d\, t, \tag{9}$$

where v_d is the drift velocity of the right edge of the void.

The atomic flux into the right edge of the void is governed by the diffusivity of the copper/barrier layer interface $D_{Cu/barrier}$, while the outgoing flux is governed by the surface diffusivity D_s. Since $D_s >> D_{Cu/barrier}$, using the Nernst-Einstein equation one can write [17]

$$v_d = \frac{eZ^*\rho j}{kT} D_s. \tag{10}$$

The EM failure occurs, when the void spans the via size, $l_{void} = L_{via}$, so that the void growth time contribution to the EM lifetime is given by

$$t_g = \frac{L_{via}}{v_d} = \frac{kT L_{via}}{eZ^*\rho j D_s}. \tag{11}$$

Electromigration in Anisotropic Metals: Solder Bump Degradation

A general three-dimensional expression for the vacancy flux $\vec{J_v}$ driven by gradients of the chemical potential and EM is given by

$$\vec{J_v} = -\frac{C_v}{kT}\mathbf{D}_v(\nabla\mu_v + |Z^*|e\nabla\varphi). \tag{12}$$

Here, C_v is the vacancy concentration, μ_v is the vacancy chemical potential, and φ is the electric potential which obeys Laplace's equation ($\Delta\varphi = 0$) [18]. Here we also introduce

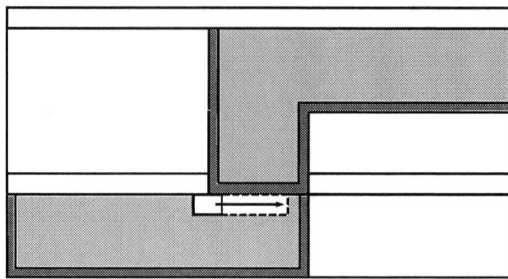

FIGURE 2. Early failure mode: slit void growth under the via.

a tensorial diffusivity $\mathbf{D}_v$, which describes the anisotropy of the vacancy transport caused by the crystal properties and the influence of mechanical deformation. The vacancy flux expression (1) and the models based on it have been widely used for the analysis of EM in dual-damascene copper interconnects. In order to model EM in solder bumps which, in addition to host atoms (e.g. tin), also include impurity atoms (e.g. nickel, copper), (1) must be extended. In our model, we assume that, prior to EM stressing, all impurity atoms in tin have occupied substitutional positions. Thus, after applying electric current, EM removes the impurity atoms from their substitutional sites and causes them to drift. Each drifting host or impurity atom induces a movement of vacancies in a direction opposite to its drifting direction and the total vacancy flux is composed of the vacancies produced by the host atoms and the impurity atoms. The total flux $\vec{J}_v^T$ is given by

$$\vec{J}_v^T = \vec{J}_v + \sum_i \vec{J}_v^i.$$ (13)

Here, $\vec{J}_v^i$ is the vacancy flux corresponding to impurity i and $\vec{J}_v$ is the flux of the host vacancies. (13) is the central equation of our model for EM-induced failure of solder bumps.

RESULTS AND DISCUSSION

The *ab initio* method described above is applied for the calculation of the effective valence inside grain boundaries and the calculated value is used to parameterize our continuum-level model [3]. Prior to carrying out the *ab initio* calculation it is necessary to construct grain boundaries with exact positions of atoms. For this purpose an in-house molecular dynamic (MD) simulator with a many-atom interatomic potential based on effective-medium theory [19] is used. The total energy of the system is expressed as

$$E_{tot} = \sum_{i=1}^{N} F(n_i) + \frac{1}{2} \sum_{i=1}^{N} \sum_{j \neq i} V(r_{ij})$$ (14)

for an N-atom system, where $V(r_{ij})$ describes a pair potential and $F(n_i)$ describes the energy due to the electron density. An example of the construction of grain boundaries by means of MD simulation is presented in Fig. 3.

104

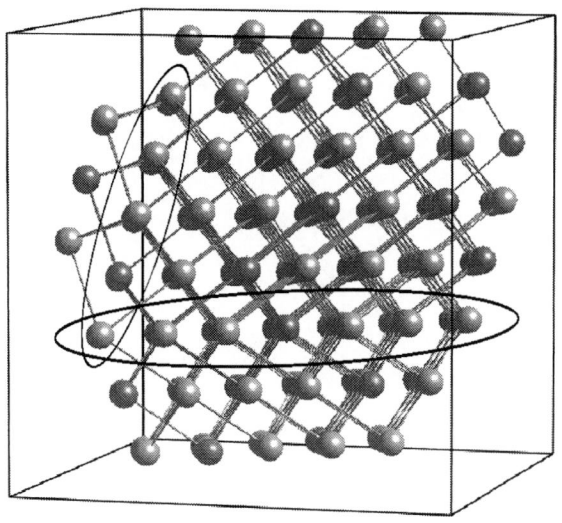

FIGURE 3. Formation of grain boundaries (circled regions).

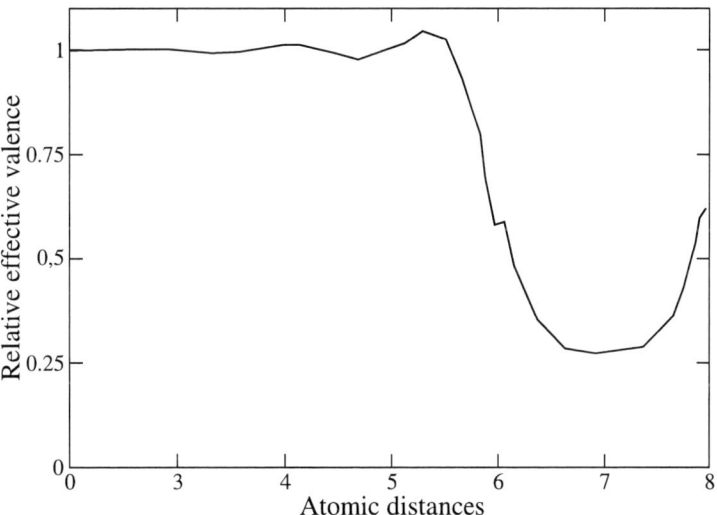

FIGURE 4. Average distribution of the effective valence near a grain boundary. The external electric field is oriented parallel to the grain boundary.

Ab initio calculations of the effective valence in copper grain boundaries have provided a value 75 % lower than in the bulk for 4.3 eV Fermi energy (cf. Fig. 4), which is in good agreement with the results of Sorbello [6]. Along with the determination of the effective valence, *ab initio* calculations predict a lowering of the energy barrier for atomistic transport. Knowing the influence of the EM force on the diffusional barrier we

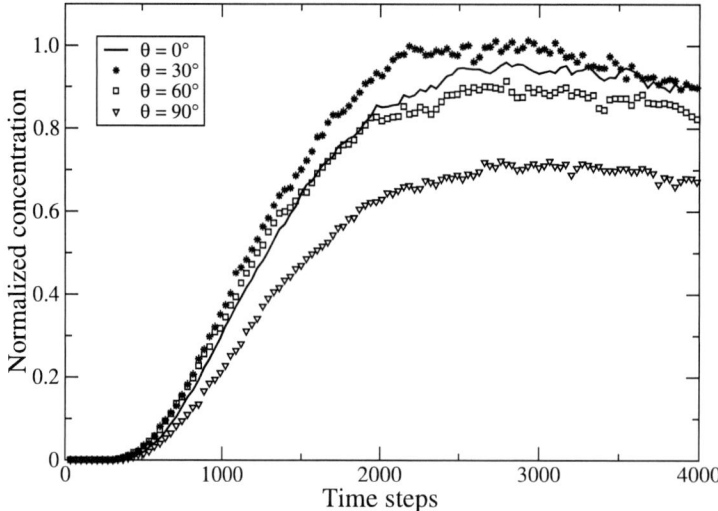

FIGURE 5. Concentration difference at four different angles (θ) between the EM force and the atom migration paths.

utilize kinetic Monte Carlo [11] simulations for EM, which provide a closer look into the distribution of atoms in the presence of EM for a specific atomistic configuration.

The dependence of the atomic concentration on the angle between the EM force and the jump direction is displayed in Fig. 5. The EM intensity clearly reduces from $\theta = 0°$, where the EM force acts in the fast diffusivity path direction, to a minimum for $\theta = 90°$, where the EM force is orthogonal to this direction.

Ab inito calculations enable us to give a proper consideration of fast diffusivity paths and microstructure in the comprehensive physically based model [3]. The solution of such a model is indeed rather complex and a detailed description of the numerical approach can be found in [15].

Fig. 6 shows the mechanical stress close to the via at the cathode end of a simulated line. A high stress develops adjacent to the via, where there is a line of intersection between the copper, the capping layer, and the barrier layer. For a copper dual-damascene M1/via structure with downstream electron flow, this is the typical site for void formation and growth leading to early EM failures.

Since EM failure has a statistical character, in order to obtain a distribution of void nucleation times several lines with different microstructures were simulated. In particular, the mechanical stress under the via was monitored for a total of twenty lines, from where the resulting stress build-up for five different structures is shown in Fig. 7.

We have observed that the time evolution of the stress curves can be divided into two main parts. In the first one the stress increases linearly with time, while in the second part it increases with the square root of time, as shown in Fig. 8 for a typical stress curve. It should be pointed out that Kirchheim [20] derived a linear stress increase from a one-dimensional version of a full physical model [3] under the condition that the stress is sufficiently low. In turn, Korhonen *et al.* [12] obtained a square root stress increase, as given by (7), from the solution of a simplified model for EM stress build-

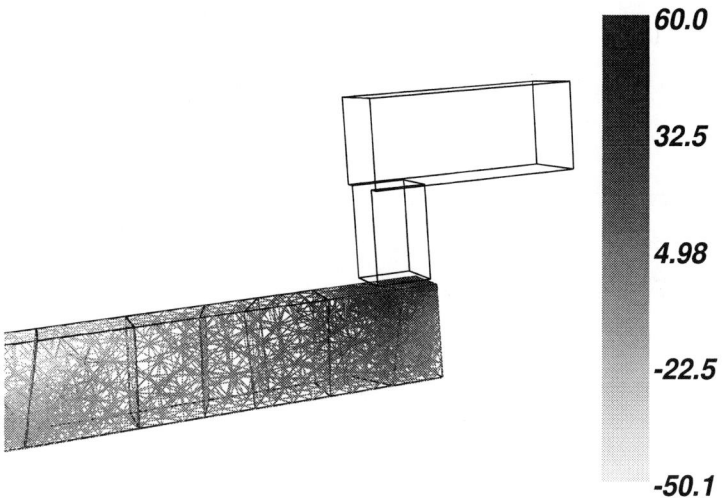

FIGURE 6. Hydrostatic stress distribution (in MPa). High stress develops at the copper/capping/barrier layer intersection adjacent to the via.

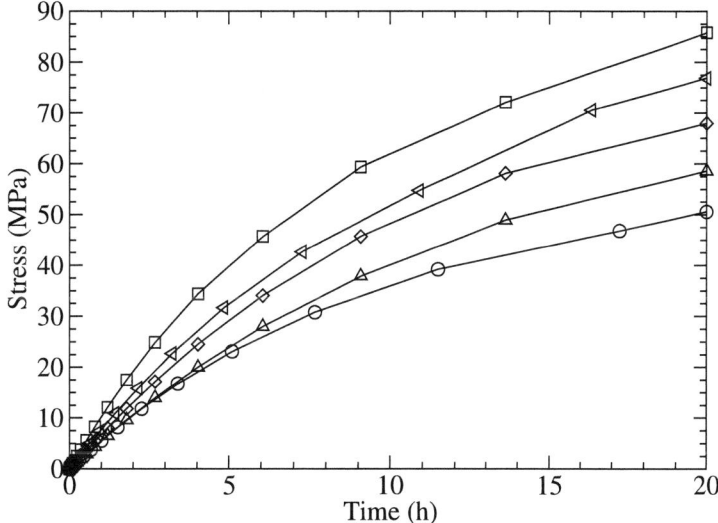

FIGURE 7. Stress build-up at the copper/capping/barrier layer intersection for lines with different microstructures.

up. Thus, the stress build-up obtained from our numerical simulations with a rather complete model and for fully three-dimensional structures can be conveniently described by simple analytical solutions.

Since void nucleation is expected to occur at high stress magnitudes, the second part of the stress curve shown in Fig. 8 is fitted by the square root model given in (7), where a is used as fitting parameter. By fitting the stress curves of all simulated structures, the distribution of the parameter a is determined, as shown in Fig. 9. The parameter is

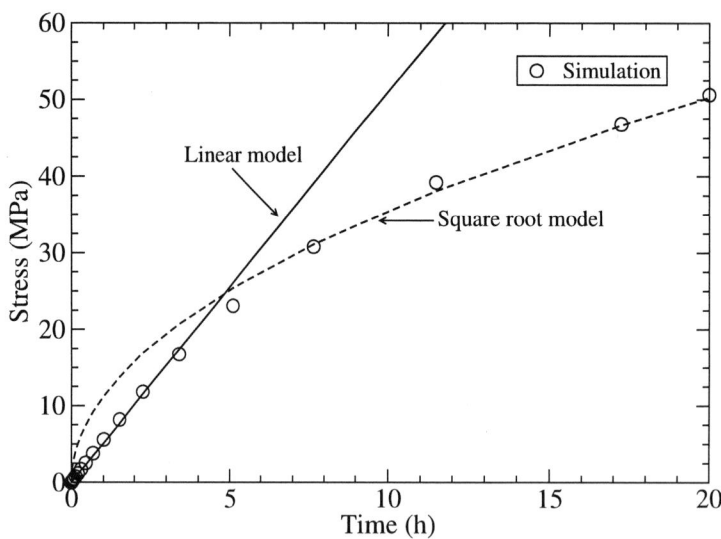

FIGURE 8. Fitting of a numerical solution using a linear and a square root model.

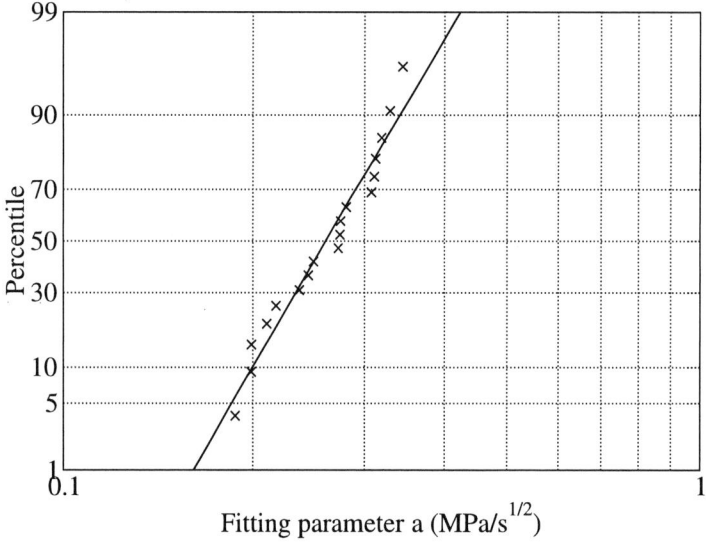

FIGURE 9. Distribution of the square root model fitting parameter. The line represents a lognormal fit.

well described by lognormal statistics, where the mean and the standard deviation are $\bar{a} = 0.23$ MPa/s$^{1/2}$ and $\sigma_a = 0.19$, respectively. Once a is known, the void formation time is obtained from (8). Since the distribution of a is also determined, we are able to obtain the statistical distribution of the void formation times, shown in Fig. 10. Due to the lognormal statistics of a, t_n also follows a lognormal distribution, where the mean and standard deviation are $\bar{t}_n = 8.5$h and $\sigma_{t_n} = 0.38$. It should be pointed out that Filippi *et al.* [21] estimated a nucleation time of approximately 5h, which lies within the range predicted by the simulations.

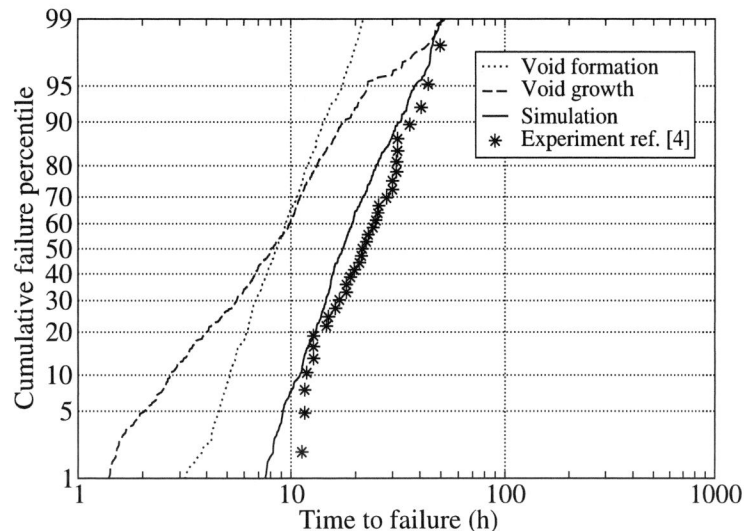

FIGURE 10. Early EM lifetime distribution.

The void growth time is determined by (11), which is a function of the surface diffusivity. Choi *et al.* [17] obtained an activation energy for surface diffusivity of 0.45 ± 0.11eV on clean copper surfaces. It is expected that their measurement delivers a more precise copper surface diffusivity than the typical ones obtained on oxidized surfaces [17] and, therefore, we have used their estimate in our simulations. Furthermore, we have assumed that the activation energy follows a normal distribution [22]. As a consequence, both the surface diffusivity and the void growth time are lognormally distributed. The mean and the standard deviation of the void growth time distribution are $\bar{t}_g = 8.0$h and $\sigma_{t_g} = 0.7$, respectively.

As the void nucleation and the void growth times are known, the early EM lifetime is given by the combination of (8) and (11),

$$t_f = \left(\frac{\sigma_c}{a}\right)^2 + \frac{kTL_{via}}{eZ^* \rho j D_s}. \tag{15}$$

The distribution of the EM lifetimes are shown in Fig. 10, together with the experimental results obtained from Filippi *et al.* [21]. The lognormal mean and standard deviation of the simulated lifetimes are $\bar{t}_f = 17.5$h and $\sigma_{t_f} = 0.41$, respectively. We can see that the simulation results provide a reasonable description for the early EM lifetimes.

A major advantage of (15) is that it is a simple analytical model which is more rigorously related to the physical mechanisms active during the early EM failure development than Black's equation. A critical issue arises, however, with regard to the estimation of the parameter a. This parameter is affected by several factors, like diffusion coefficients, effective valence, mechanical moduli, microstructure, etc, so that it cannot be defined in a closed form as in full physicall modeling [3],[15]. Nevertheless, we have observed that it can be related to Korhonen's solution. In this way, it can be directly described by an analytical expression and connected to physical parameters according to (7).

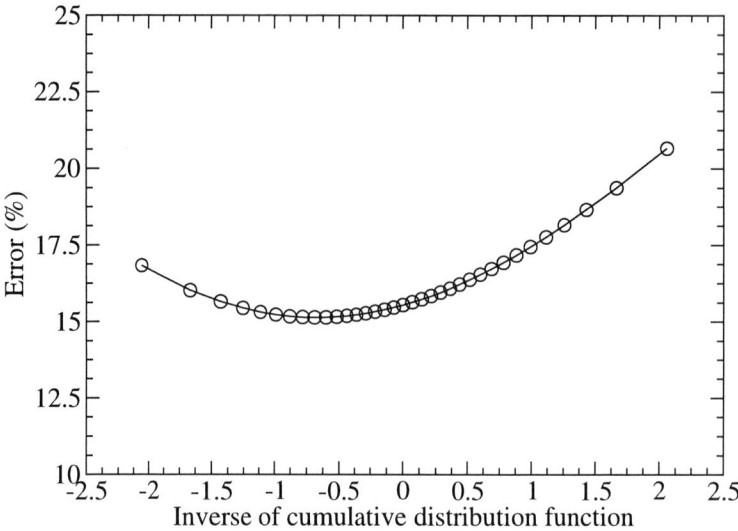

FIGURE 11. Error between the simulation and the experimental results.

The relative difference between the simulated and experimental lifetimes for the same failure percentile varies between 15% and 20%, as shown in Fig. 11. The difference is smaller for shorter lifetimes, since the proposed slit void growth model is more accurate for very early failures, where the void volumes are smaller. Such an error magnitude is reasonable, given the required assumptions for the parameters and considering the simplicity of the model.

In the final example we discuss EM failure in tin solder bumps, which is influenced by the crystal orientation of grains. During EM testing two failure modes are observed [23]:

- Mode 1. The cathode nickel barrier layer and the inter-metallic compound (IMC) remain intact, while EM induced voids are formed at the tin solder bump interface to the IMC.
- Mode 2. The nickel barrier layer and the IMC are depleted and swept away. Inside the nickel layer a void is formed.

The interconnect structure used for the simulation study is presented in Fig. 12.

This structure consists of a tin based solder bump with nickel under bump metallization (UBM) at the cathode end. The cathode and anode ends are connected to copper interconnect layers. In order to study the complete problem we must consider EM in the nickel and the tin segments as well as at their interface. As we know from experimental observations, at the interface between nickel and tin, an IMC is formed. EM in the nickel and tin segments is described by the standard EM model [18].

For the IMC we assume at first a simple nickel segregation model [24]. Nickel has a face-centered cubic crystal structure and in the unstressed state self-diffusion and EM in such crystal structures is isotropic.

In this work the predominant focus is on the effects of EM; therefore, the models for

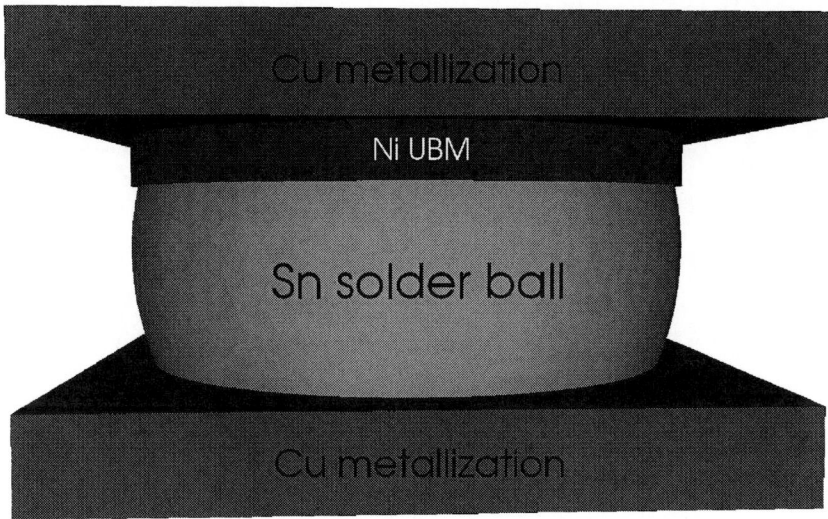

FIGURE 12. Solder bump structure used for simulation.

stress induced anisotropy of the diffusivity tensor are neglected [25]. First we study the situation where the c-axis of the tin grains exibits a large angle with the current flow direction, where the rate of nickel diffusion in tin is small. Failure is mainly due to tin self-EM, resulting in a peak vacancy concentration between the IMC and the solder. This case corresponds to the failure Mode 1.

The failure Mode 2 is obtained by rotating the tin crystal by 90° when the crystal c-axis becomes aligned with the electric current flow direction. Nickel atoms are transported from the UBM layer through the IMC into the solder bump below, where they electromigrate rapidly along the c-axis. At this instance, the peak in vacancy concentration occurs in the UBM nickel layer. The locations of vacancy concentration peaks in both failure modes correspond to sites of peaks of tensile stress, which are sites of void nucleation.

In Fig. 13 we compare the time dependent rise in the vacancy concentration due to tin self-diffusion and self-EM for different crystal orientations. Three cases are studied:

- c-axis parallel to the current flow direction
- c-axis rotated by $\theta = 30°$
- c-axis rotated by $\theta = 60°$

The impact of IMC and nickel-related vacancy-influx is neglected in order to obtain a clear picture of the influence of the crystal anisotropy on material transport.

As we can see in Fig. 13, a crystal rotation causes a reduction of the EM intensity. We denote with θ an angle between the current flow direction and the c-axis of the crystal. By comparing the curves for $\theta = 0°$ and $\theta = 60°$ we see that the vacancy concentration at $\theta = 0°$ keeps rising, while the curve at $\theta = 60°$ stays almost at equilibrium. This result helps to understand the experimental observation of failure development in solder

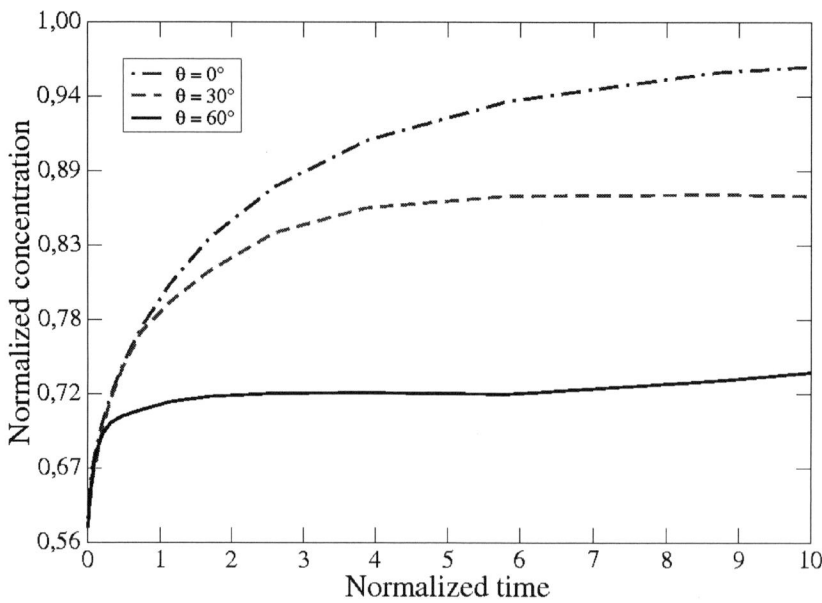

FIGURE 13. Increase in vacancy concentration for three different crystal orientations.

bumps which consist of two large grains. In these cases the grain with the *c*-axis oriented in parallel to the current flow is completely swept away, while the other grain is still intact [5].

CONCLUSION

Our work demonstrates a novel approach for the calculation of the EM force on an atomistic level and its application to continuum-level modeling. The consideration of the accurate effective valence in grain boundaries enables a realistic simulation of EM behavior. The presented combination of atomistic force calculations with a kinetic Monte Carlo simulation enables sophisticated analyses of vacancy dynamics. A compact model for estimation of the early EM lifetimes in M1/via structures of copper dual-damascene interconnects was developed. The model was derived through the combination of a complete model for void nucleation together with a simple slit void growth mechanism under the via. Given the simplifications and assumptions made for the simulations, a reasonable approximation to experimental early EM failures has been obtained. Finally, a physical model of EM in anisotropic crystalline tin has been presented and used as a basis for the investigation of the failure behavior for a realistic solder bump structure by means of simulation.

REFERENCES

1. Z.-S. Choi, R. Mönig, and C. V. Thompson, *Appl. Phys. Lett.*, vol. 90, p. 241913, (2007).
2. E. Zschech and P. R. Besser, *Proc. Interconnect Technol. Conf*, pp. 233-330, (2000).
3. H. Ceric, R. L. de Orio, J. Cervenka, and S. Selberherr, *IEEE Trans. Dev. Mat.Rel.*, vol. 9, pp. 9, (2009).
4. T. R. Bieler, H Jiang,L. P. Lehman, T. Kirkpatrick, E. J. Cotts, and B. Nandagopa, *IEEE Trans. Comp. Pack. Techn.*, vol. 31, nr. 2, pp. 370-381, (2008).
5. C. Y. Liu, C. Chen, C. N. Liao, and K. N. Tu, *Appl. Phys. Lett.*, vol. 75, no. 1, pp. 58-60, (1999).
6. R. S. Sorbello, in *Materials Reliability Issues in Microelectronics*, edited by J. R. Lloyd, F. G. Yost, and P. S. Ho, vol. 225 pp. 3-10, (1996).
7. R. P. Gupta, *Phys. Rev. B*, vol. 25, pp. 5118-5196, (1982).
8. D. N. Bly and P. J. Rous, *Phys. Rev. B*, vol. 53, pp. 13909, (2006).
9. H. Ceric, R. L. de Orio, F. Schanovsky, W. Zisser, and S. Selberherr, *Proceedings of the 16th International Conference on Simulation of Semiconductor Processes and Devices*, pp. 135-138, (2011).
10. G. Kresse and J. Furthmüller, *Phys. Rev. B*, vol. 54, pp. 11169, (1996).
11. R. Sorensen, Y. Mishin, and A. F. Voter, *Phys. Rev. B*, vol. 62, pp. 3658, (2000).
12. M. A. Korhonen, P. Borgesen, K. N. Tu, and C.-Y. Li, *J. Appl. Phys.*, vol. 73, no. 8, pp. 3790-3799, (1993).
13. R. J. Gleixner, B. M. Clemens, and W. D. Nix, *J. Mater. Res.*, vol. 12, pp. 2081-2090, (1997).
14. M. W. Lane, E. G. Liniger, and J. R. Lloyd, *J. Appl. Phys.*, vol. 93, no. 3, pp. 1417-1421, (2003).
15. R. L. de Orio, Dissertation, Technische Universität Wien, (2010). [Online]. Available: http://www.iue.tuwien.ac.at/phd/orio/
16. A. S. Oates and M. H. Lin, *IEEE Trans. Device Mater. Rel.*, vol. 9, no. 2, pp. 244-254, (2009).
17. Z. S. Choi, R. Mönig, and C. V. Thompson, *J. Appl. Phys.*, vol. 102, p. 083509, (2007).
18. H. Ceric, R. Heinzl, Ch. Hollauer, T. Grasser, and S. Selberherr, *Stress-Induced Phenomena in Metallization, AIP*, pp. 262-268, (2006).
19. K. W. Jacobsen, J. K. Norskow, and M. J. Puska, *Phys. Rev. B*, vol. 35, pp. 7423, (1987).
20. R. Kirchheim, *Acta Metall. Mater.*, vol. 40, no. 2, pp. 309-323, (1992).
21. R. G. Filippi, P.-C. Wang, A. Brendler, P. S. McLaughlin, J. Poulin, B. Redder, and J. R. Lloyd, *Proc. Intl. Reliability Physics Symp.*, pp. 444-451, (2009).
22. L. Doyen, X. Federspiel, L. Arnaud, F. Terrier, Y. Wouters, and V. Girault, *Proc. Intl. Integrated Reliability Workshop*, pp. 74-78, (2007).
23. M. Lu, *Stress-Induced Phenomena in Metallization, AIP*, pp. 229-234, (2012).
24. F. Lau, L. Mader, C. Mazure, C. Werner, and M. Orlowski, *Appl. Phys. A*, vol. 49, pp. 671-675, (1989).
25. R. L. de Orio, H. Ceric, and S. Selberherr, *J. Comp. Electronics*, vol. 7, no. 3, pp. 128-131, (2008).

Physics-Based Simulation of EM and SM in TSV-Based 3D IC Structures

Armen Kteyan[1], Valeriy Sukharev[2], and Ehrenfried Zschech[3]

[1]*Mentor Graphics Corp., 0012 Yerevan, Armenia*
[2]*Mentor Graphics Corp., Fremont, CA 94538 USA*
[3]*Fraunhofer Institute for Nondestructive Testing IZFP, D-01109 Dresden, Germany*

Abstract. Evolution of stresses in through-silicon-vias (TSVs) and in the TSV landing pad due to the stress migration (SM) and electromigration (EM) phenomena are considered. It is shown that an initial stress distribution existing in a TSV depends on its architecture and copper fill technology. We demonstrate that in the case of proper copper annealing the SM-induced redistribution of atoms results in uniform distributions of the hydrostatic stress and concentration of vacancies along each segment. In this case, applied EM stressing generates atom migration that is characterized by kinetics depending on the preexisting equilibrium concentration of vacancies. Stress-induced voiding in TSV is considered. EM induced voiding in TSV landing pad is analyzed in details.

Keywords: stress migration, electromigration, TSV, vacancy, FEA simulation.
PACS: 66.30.Fq, 66.30.Qa

INTRODUCTION

Large stresses generated by TSVs, which were introduced in novel 3D IC stacking technologies for providing the inter-die interconnection, cause the serious concerns regarding chip performance and reliability [1]. Process-induced stresses generated inside TSVs have a notable impact on the performance of nearby transistors [1]. So-called "keep-out" zones surrounding TSVs were introduced in order to prevent the transistors placing in highly stressed regions. Among the reliability issues caused by TSV we can mention origination of interfacial cracking or delamination at the TSV sidewalls in the vicinity of silicon/interconnect interface, and generation of the cohesive cracks in silicon and interconnect dielectric [2].

Initial process-induced stresses inside TSV are characterized by non-uniform distribution along its height. The resulting stress gradients tend to redistribute atoms and to establish a uniform stress distribution. The main sources of internal stresses in TSVs are: a grain growth occurring during microstructure evolution, and a mismatch of thermo-mechanical properties of TSVs and surroundings. The post plating growth of copper grains results in merging of the small initial grains and forming the larger grains. This causes shrinkage of the initial volume and results in a tensile strain due to interaction with the rigid confinement. Thermal stress is generated by cooling the chip down from the stress-free anneal state to the test/use condition due to the mismatch of the coefficients of thermal expansion of the involved materials such as copper, silicon, composite interconnect, etc. Additional stresses are generated by many other steps of the 3D IC stack fabrication such as dies mounting, solder balls solidification, etc. These processes can result in a sequence of die warpages that generates a "global stress" distributed across the die. All these stresses acting together may be important for accurate description of EM-induced void nucleation [3].

Despite of the existing solid understanding of SM/EM phenomena in copper interconnects new challenges arise due to development of 3D packaging technologies. One reason is the strong elastic interaction between TSVs and the neighboring interconnect metal lines, caused by a huge amount of metal contained in a single TSV, which is responsible for generating the big deformations in the surroundings. Different types of TSV's architectures and different copper-fill technologies are responsible for different configurations of generated stress fields surrounding TSVs. The latter can be responsible for different types of damages generated by these stresses. Another problem is related to the employment of new materials in 3D IC technology, for example the novel alloys used for solder balls fabrication. Finally, the complex geometries of many components of 3D ICs interconnect architecture – TSV, landing pad, back-end of line (BEOL) interconnect, back side redistribution layer (BRDL), may require an employment of the detailed physics-based finite element analysis (FEA) simulations for the stress evolution study.

In this paper we describe a methodology for the simulation-based analysis of SM and EM-induced evolution of the stresses in TSVs. The developed model is used to describe the stress-induced voiding in a TSV and in the landing pad in BRDL metallization.

PHYSICAL MODEL OF STRESS EVOLUTION

A comprehensive physical model of stress evolution (see [4-5] and the references provided there) is employed for analysis of the 3D package interconnect. The gradients of hydrostatic stress σ_{Hyd} and vacancy concentration N, together with the electron wind force (if an electric load is applied) are responsible for the vacancy flow origination. Evolution of the vacancy concentration is described by the continuity equations, which have different forms for the grain bulk, and for interfaces and grain boundaries (GB), which are the paths for fast vacancy migration and the sites for vacancy generation/annihilation (everywhere further the index "I" corresponds to the properties of interfaces and GBs).

$$\frac{\partial N}{\partial t} + \vec{\nabla}(-D\vec{\nabla}N - \frac{DN}{kT}((1-f)\Omega\vec{\nabla}\sigma_{Hyd} + eZ\vec{\nabla}V)) = 0 \tag{1}$$

$$\frac{\partial N}{\partial t} + \vec{\nabla}(-D_I\vec{\nabla}N - \frac{D_I N}{kT}((1-f)\Omega\vec{\nabla}\sigma_{Hyd}^I + eZ_I\vec{\nabla}V)) + G = 0 \tag{2}$$

The generation-annihilation term is

$$G = -\frac{N - N_I^{eq}}{\tau} \tag{3}$$

where

$$N_I^{eq} = N_0 \exp\left(f\Omega\sigma_{Hyd}^I / kT\right) \tag{4}$$

is the vacancy concentration equilibrated with the stress, N_0 is the vacancy concentration in the zero stress condition, Ω is the atomic volume, e is the electron charge, eZ and eZ_I are the effective charges of the migrating atoms in the bulk and on interfaces and GBs, respectively. V is the local electrical potential, k is Boltzmann's constant, and T is the absolute temperature; D and D_I are the diffusivities of vacancies in the bulk and on the interfaces and GBs. In (4) $f = \Omega_V / \Omega$ is a ratio of the lattice volume occupied by a vacancy (Ω_V) to the volume occupied by an atom, which, as it follows from the quantum chemical estimations, has a value approximately equals to 0.6 [6]. A presence of the multiplier f in (4) can be explained by the work done against the hydrostatic stress during the vacancy formation. It includes the work $\Omega\sigma_{Hyd}$ done against pressure when a transferred lattice atom deforms the GB or interface, and the work performed during the volume relaxation around the newly formed vacancy: $-(1-f)\Omega\sigma_{Hyd}$. A combined effect $f\Omega\sigma_{Hyd}$ is responsible for the reduction of the energy of vacancy formation by the applied hydrostatic tension. Note that a replacement of the normal stress component, used by Herring [7] for calculating the work of GB deformation, by σ_{Hyd}, was done here to easy the simulation setup; however, this should not introduce a big error in final results. Here and everywhere below the vacancy concentration is a fraction of empty lattice nodes in the unit volume of the crystal.

Model assumes that vacancy diffusivity is characterized by the Arrhenius type of dependency on temperature

$$D = D_0 \exp\left\{-\frac{E_A - \Omega^*\sigma_{Hyd}}{kT}\right\} \tag{5}$$

Here, E_A is the effective activation energy of the vacancy diffusion, which is different for diffusion through the bulk, along interfaces, and GBs [8-10]. The activation volume $\Omega^* \approx 0.95\Omega$ is a combination of the formation volume, which is the crystal volume change upon formation of a vacancy in its standard state, and the migration volume, which is the additional volume change when the defect reaches the saddle point in its migration path [11].

Both mechanisms affecting the vacancy distributions, which are the vacancy diffusion and their generation/annihilation are essential in stress evolution kinetics. An elemental act of vacancy generation means that an atom leaves its position in the lattice and (most probably) is placed at an interface or GB site. These plated atoms are assumed to be immobile, so they exist at the GBs and interfaces and their concentration (M) is determined only by the generation terms:

$$\frac{\partial M}{\partial t} + G = 0 \tag{6}$$

Plated atoms generate an additional compressive stress, which can be described by introducing or removing an extra volume Ω on the GBs and interfaces. Hence, the total volume strain during the vacancy generation/annihilation is determined by the deviation of the concentrations of vacancies and plated atoms from the zero-stress values, N_0 and M_0:

$$e_V = -(1-f)(N-N_0) + (M-M_0) \tag{7}$$

Concentrations of vacancies as well as plating atoms are expressed in $1/\Omega$ units, and they represent the local fraction of lattice sites occupied by the vacancies or plating atoms.

The total generated strain tensor ε_{ij} includes inelastic volume deformation ε_{ij}^{inel}, due to change of concentration of vacancies (ΔN) and plated atoms (ΔM), and the elastic component ε_{ij}^{el}, which is due to interaction with the confinement [12]:

$$\varepsilon_{ij} = \varepsilon_{ij}^{el} + \varepsilon_{ij}^{inel},$$
$$\varepsilon_{ij}^{inel} = \frac{1}{3}\left(-(1-f)\Delta N + \Delta M\right)\delta_{ij} \tag{8}$$

Hooke's law relates the generated stress with elastic strain through the stiffness matrix $\{C_{ij,kl}\}$:

$$\sigma_{kl} = C_{ij,kl}\varepsilon_{ij}^{el} \tag{9}$$

This enables to determine the total hydrostatic stress:

$$\sigma_{Hyd} = C_{ij,kk}\left[\varepsilon_{ij} + \frac{1}{3}\left((1-f)\Delta N - \Delta M\right)\delta_{ij}\right] \tag{10}$$

Combining equations (1) - (10) with the force balance condition [13]

$$\frac{\partial \sigma_{kl}}{\partial x_k} = 0 \tag{11}$$

we can obtain the distribution of stresses, using the finite element analysis.

The presented formulation of stress evolution enables us to describe the effect of texture of studied segments as well as the effect of copper anisotropy on the stress evolution. If the orientations of grains regarding the studied segment direction are known, then the stiffness matrix for each grain can be transformed by means of rotation tensors. This approach was successfully employed for descriptions of the EM-induced voiding in interconnect wires with a bamboo structure [14]. Performed simulations have revealed a non-trivial distribution of interfacial shear stresses at the near cathode region that was originated by a specific orientation of neighbor grains. Peaks in the simulated shear stress distribution were located at sites where void nucleation was observed in the experiment. It was concluded that the shear stress is responsible for the crack formation.

Similar detailed simulation of the texture related effects cannot be employed for the analysis of the state of stress inside TSV: numerous small grains existing in large TSVs cannot be modeled individually. Therefore, the main microstructure related effects which are considered in this report are the initial distribution of growth stress along the TSV and the microstructure related distribution of vacancies diffusivities.

MICROSTRUCTURE RELATED STRESS GENERATION AND EVOLUTION

Grain growth stress is generated in the TSVs interior upon completion of the copper plating. Initial, as deposited, small Cu grains merge during the annealing/self-annealing processes, resulting in the grain growth and, as a sequence, the disappearance of a part of GBs. If the initial size of grains is ~d_0 and the final size is ~d, than the generated growth strain can be described by the following expression:

$$\varepsilon_{gr}(z) = \delta\left(\frac{1}{d_0} - \frac{1}{d(z)}\right) \tag{12}$$

Here, z is the coordinate along the TSV, and the parameter $\delta \sim 1 A^\circ$ is the "thickness" of a grain boundary. This description intends to take into account the non-uniform distribution of post-anneal grain sizes. The regions with the larger grains are characterized by the larger growth strain.

A dependence of the grain size distribution on the recipe of copper electroplating process was studied in [15]. It was shown that in the conventional deposition process, which is characterized by formation of a thick overburden layer, grain growth at the top of TSV is affected by the overburden layer, where large grain formation occurs. As a result, grains formed at the via top exceed the size of grains formed in the middle and bottom regions of TSV. On the other hand, growth of the grains at the TSV bottom is restricted by the bottom plane and by the sidewalls. As a result, the generated strain is distributed as: ε_{gr}(top) > ε_{gr}(middle) > ε_{gr}(bottom). Different distribution of the grain sizes along the TSV height is observed in the case of a specially tuned electroplating process, when the segregation of levelers at the top of via opening prevents grain growth in this region. In this case, the strain distribution is: ε_{gr}(middle) > ε_{gr}(bottom) > ε_{gr}(top). Growth stress can be estimated assuming that as-deposited grain sizes are distributed in the interval of $d_0 \sim$ 30-50 nm, while the sizes of grown grains can vary from 100 to 3000 nm [15]. Assuming the Young's modulus for silicon and copper as $E_{Si} \sim$170 GPa and $E_{Cu} \sim$120 GPa, we can estimate the values of generated growth stresses as ~$200 - 400$ MPa.

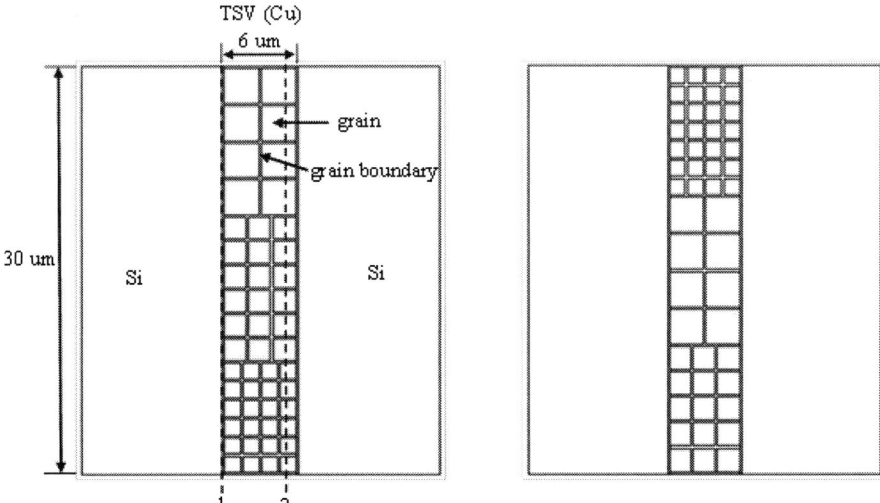

FIGURE 1. Simulation domain: a) large Cu grains at the top of TSV, b) large grains in the middle region. Evolution of stress distribution along the lines 1 and 2 is studied.

117

The schematics of the grain size distribution in the mentioned two cases are demonstrated in Fig.1. Fixed boundaries of the studied domains and continuity of displacements and normal stresses on the Si – TSV interfaces are accepted as boundary conditions in the simulation analysis described below. TSVs of 6um in diameter and 30um in height are considered.

In both studied cases the stress gradients create diffusional flow of copper atoms in accordance with the vacancy transport mechanism. For a simple one-dimensional case, corresponding to TSV interface, evolution of vacancy concentration during SM is described by Eq.(2), where the EM-related term is vanished. Analytical solution of SM equations shows that the final steady-state is characterized by the uniform hydrostatic stress distribution along TSV, which is just the average of initial growth stress [16]:

$$\sigma_{Hyd}^{eq} = \bar{\sigma}_0 \equiv \frac{1}{L} \int_L \sigma_{Hyd}^0 (z) dz$$

$$N^{eq} = N_0 \exp\left\{ \frac{\sigma_{Hyd}^{eq}\Omega}{kT} \right\}$$

$$M^{eq}(y) = \frac{1}{B}\left(\sigma_{Hyd}^0(y) - \sigma_{Hyd}^{eq} \right)$$

(13)

where B is the corresponding bulk modulus. This conclusion was proved by simulation, performed with the commercial FEA tool COMSOL Multiphysics [17], of the steady state and transient distributions of stress components and concentrations of vacancies and plated atoms. These evolutions are demonstrated in Fig.2 for two cases characterized by different initial distributions of stress, that were created by different post-anneal distributions of the grain sizes along the TSV height. Fig. 2,a describes these distributions in the case where large grains were formed at the TSV top, and Fig. 2,b, in the case where large grains were formed in the middle part of the TSV. In the both cases the uniform steady-state distributions of the stress and vacancy concentration were reached at the interface (along the line 1 in Fig.1). Steady-state profile of the distribution of concentration of plated atom corresponds to the initial stress distribution. It can be understand as following: in the regions of high initial (tensile) stress the equilibration of σ_{Hyd} is achieving by the increase of concentration of plated atoms M, while, in the less stressed regions the decrease of M concentration is needed. It should be noted, that in the simulations, as well as in the analytic analysis resulted in equation (13), the initial concentration of plated atoms $M_0=0$ was assumed. It results that the decreased concentration of plated atoms is described by negative M values.

As it was mentioned above, interfaces and GBs were considered as the venues for fast vacancy diffusion. It was accounted by introduction of the enhanced vacancy diffusivities along GBs and interfaces as $D_{GB}=10^2 D$, $D_{int}=10^3 D$, where D is the vacancy diffusivity in the grain interior. The steady state distributions of all involved characteristics were achieved at t ~ 10^5 arbitrary units (a.u.). Obviously, a longer time will be required to establish the steady state in the case of deeper TSVs.

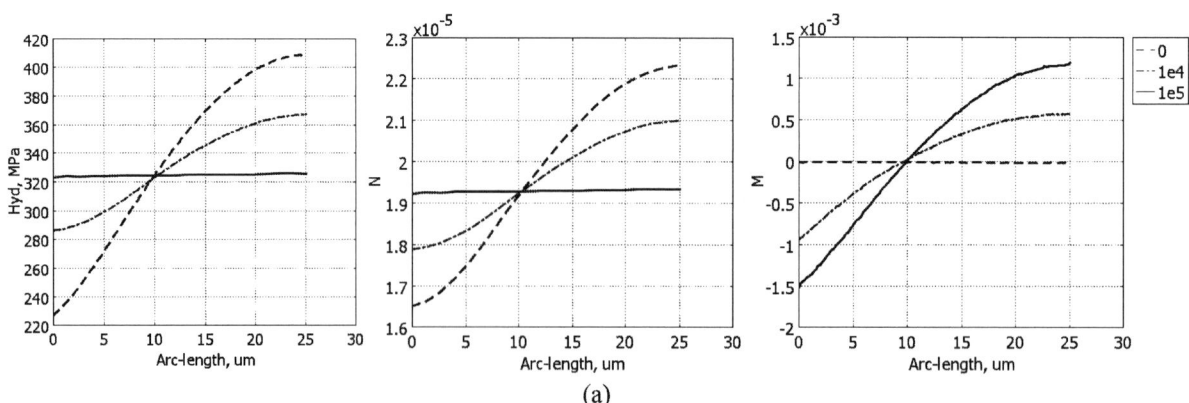

(a)

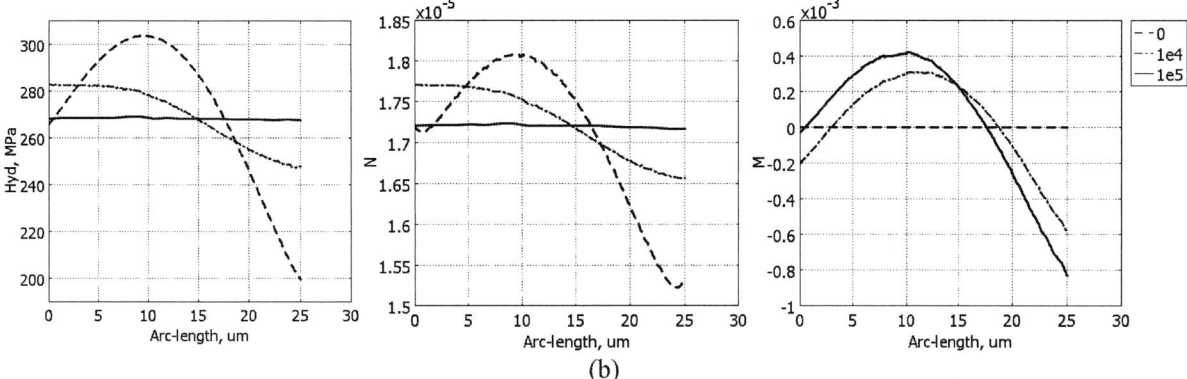

(b)

FIGURE 2. Time evolution of the interface distributions of hydrostatic stress, and concentrations of vacancies and plated atoms simulated for different distributions of the initial hydrostatic stress: (a) monotonically increase from the TSV bottom to top, (b) maximal initial hydrostatic stress in the TSV center. The distributions correspond to 0, 10^4, and 10^5 a.u. of time.

Different results were obtained for the grain interior regions where the generation/annihilation of the vacancy/plated atom pairs did not take place. Calculated distributions of hydrostatic stress and vacancy concentration, along the line 2 in Fig.1, are presented in Fig.3 for the case of the "large grains on the TSV top". The distributions of the tips of spikes existing on the simulated curves, which correspond to the intersections of the line 2 with GBs, precisely fit to the distributions of stress along interface shown in Fig.2,a. Inside the grains some stress relaxation took place, but the gradients of stress and vacancy concentration were not vanished.

While stress evolution at the interfaces and GBs is caused mainly by plated atoms by means of the dilatational effect and the elastic load from the confinement, the evolution of stress inside grains is caused by dilatation generated by vacancies. Grain interiors, being incapable to generate vacancies, are receiving them from or injecting them into the interfaces/GBs by means of slow diffusion. Sourcing or draining the vacancies by grain interior is responsible for generation or annihilation of the new pairs of these point defects (vacancy/plated atom) at the interfaces/GBs.

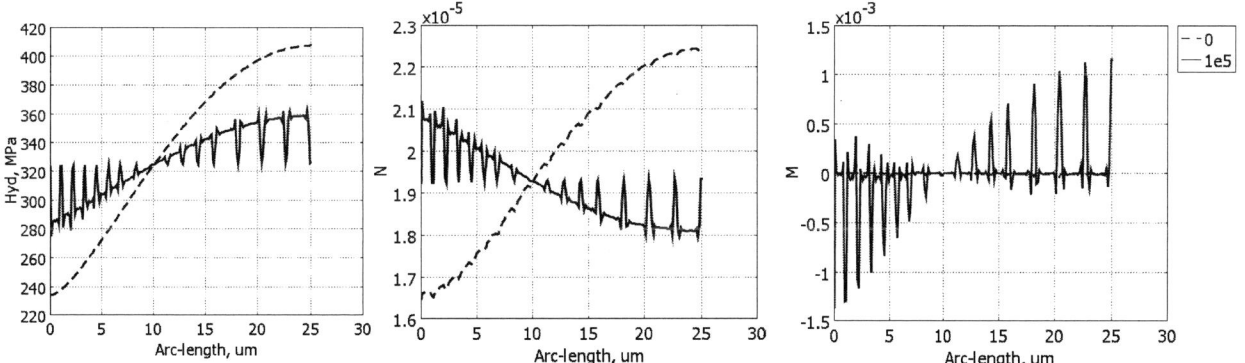

FIGURE 3. Evolution of the distributions of hydrostatic stress (a), concentrations of vacancies (b) and plated atoms (c) inside grain regions (along the line 2 in Fig.1).

The generation/annihilation of the pairs "vacancy - plated atom", accompanied by the migration of vacancies, lasts till a steady state condition will be achieved. It is characterized by the vanishing the total flux of vacancies caused by the stress and concentration gradients, Fig.4:

$$\nabla N + N \frac{\nabla \sigma_{Hyd} \Omega}{kT} = 0 \tag{14}$$

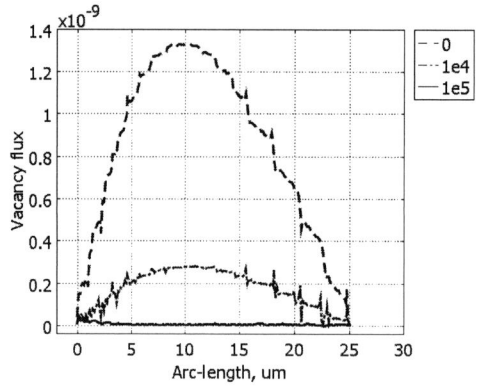

FIGURE 4. Vacancy flux evolution inside TSV.

EFFECT OF TSV ARCHITECTURE ON STRESS EVOLUTION KINETICS

An accurate description of stress evolution inside TSVs requires accounting the detailed geometry of the surrounding confinement. Two different types of TSV architecture are considered in this paper. The "via-last" TSV, which is processed when all layers of the BEOL interconnect have been processed already, is shown in Fig.5,a. Fig.5,b shows a "via-middle" TSV, which is processed just after completion of a contact layer. The main difference in the architectures of these TSV types is different neighborhoods around the TSV top. Via-last is landed at the relatively thick and rigid pad in RDL, while its top portion of sidewalls is surrounded by the soft BEOL. Via-middle is surrounded by Si everywhere along the sidewalls, and it is attached to the thin M1 layer of BEOL interconnect. The mechanical properties of the involved materials were determined by Young's modulus values:

TSV, RDL/pad: copper, E_{Cu}=120 GPa;

Substrate: silicon, E_{Si}=170 GPa;

Interconnect (BEOL): composite, E_{int}=30 GPa.

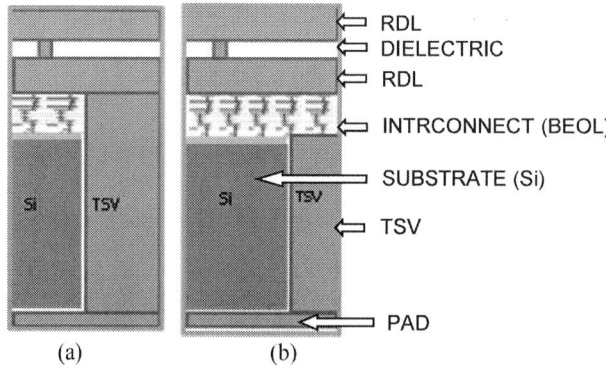

FIGURE 5. Schematics of the TSV constructions: via-last (a), via-middle (b).

As it was mentioned above, a large number of grains inside TSV makes a detailed description of the copper-fill texture almost impossible in the frame of FEA simulations. A continuum description of the major characteristics such as distributions of the grain sizes and the generation/annihilation rates along the TSV height, governing the stress evolution should be adopted. Everywhere further we prescribe a larger diffusivity and a larger generation/annihilation rate to the region with the smaller grains due to a large net volume of GBs.

The aim of the simulations is to demonstrate the effect of a thermal treatment of the studied 3D ICs on the stress evolution inside the TSV, prior the electrical loading. This "thermal treatment" procedure is introduced to model the stress relaxation in the TSV during different high-temperature process steps. Three subsequent stress relaxation steps caused by vacancy/plating atom generation/annihilation and vacancy migration were simulated: (i) initial "growth stress" relaxation at $T_{treatment}$ of 428K; (ii) stress relaxation after cooling down to the T_{test} of 373K; and (iii) stress evolution after an electrical load (voltage) was applied at T_{test}. Evolutions of stress distributions obtained for these steps are demonstrated in Fig.6 for via-last TSV in the case of the "large grains are on the TSV top". An abrupt change of the monotonic stress increase, taking place while moving from the TSV bottom toward its top, which takes place in the vicinity of the silicon/interconnect interface is caused by the change of the confinement from the rigid silicon to the soft BEOL interconnect. This distribution becomes more uniform due to stress gradient induced migration taking place during the thermal processing. Sufficiently long treatment results in the stress relaxed to a constant value (Fig.6,a). The consequent cooling down to the test temperature increases the tensile stress and introduces new non-uniformity (Fig.6,b). At this step the full equilibration of the stress by means of redistribution of lattice defects takes longer time, due to the lower temperature. If the electric load is applied when the full relaxation of this thermal stress is achieved, then the EM-induced stress evolution has the well-established behavior, which results in the linear distribution in the steady-state, with the slope determined by the applied current density (Fig.6,c).

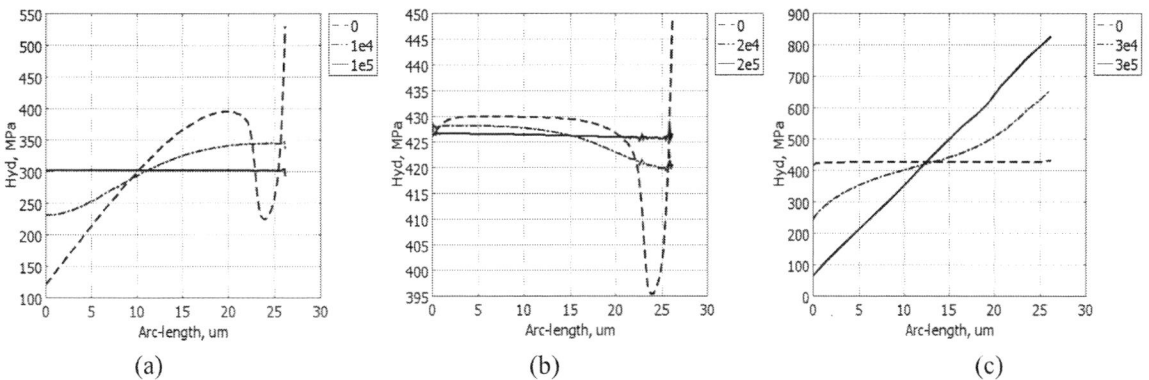

(a)   (b)   (c)

FIGURE 6. Hydrostatic stress distributions in the via-last TSV interface in the case of large grains on the top, at the states of equilibrium achieved after anneal-induced microstructure transformation (a), after cooling down to the test temperature (b), and after applying a DC voltage (c).

Initial distribution of growth stress in the "via-middle" case is characterized by decrease of the stress near the top of TSV, which is attached to the soft interconnect. The evolution of stress distributions with time, presented in Fig.7, is similar to one obtained in the "via-last" case.

A major problem with predicting the EM-induced failures might happen when the non-uniform stress distribution was not fully relaxed prior to electric load. In this case the initial stress distributions must affect the kinetics of EM-induced void nucleation [16]. Therefore, the information on anneal temperature and its duration is necessary for proper modeling of the EM-induced voiding.

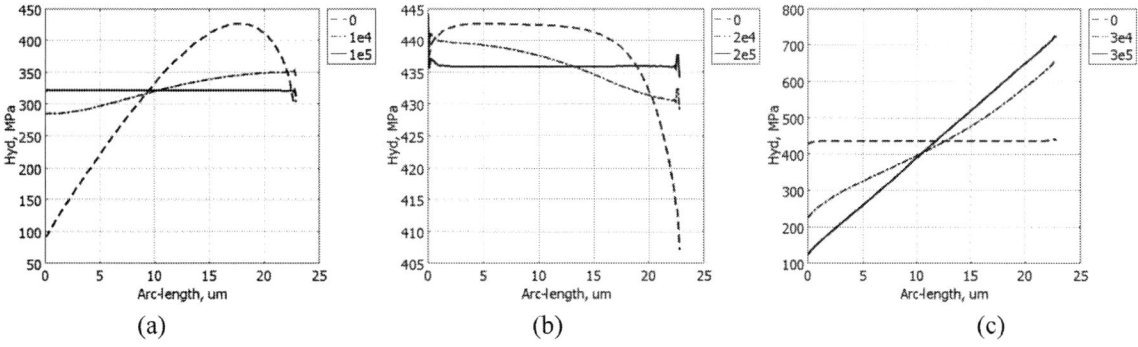

(a)   (b)   (c)

FIGURE 7. The same distributions as in Fig.6, in the case of the via-middle TSV.

STRESS INDUCED VOIDING IN TSV

All discussed above results of the stress evolution inside TSVs were obtained under assumption that no voids were generated. If opposite is valid then the further, post-voiding stress relaxation is governed by the void growth. Stress-free surface of the void creates additional stress gradients, which drive vacancies toward the void, resulting in the void growth. This process is accompanied by a decrease of the existing tensile stresses. We can estimate a volume of the saturated void in TSV as the volume of the void, which is required for disappearance of the growth and thermal strains [18]:

$$V_{void} = \left(\bar{\varepsilon}_{gr} + \bar{\varepsilon}_{therm} \right) V_{TSV} \tag{15}$$

Taking into account the values of initial and post-anneal grain sizes, which were used in the previous sections, and a mismatch of the coefficients of thermal expansion for copper and silicon $\Delta\alpha \approx 14 \cdot 10^{-6}$ K^{-1}, we can obtain the average values of the growth and thermal strains: $\bar{\varepsilon}_{gr} \sim 5 \cdot 10^{-3}, \bar{\varepsilon}_{therm} \sim 4 \cdot 10^{-3}$. Complete relaxation of these strains inside TSVs with a diameter of 6 um and height of 30 um requires generation of the spherical void of 2.4 um diameter. Such void is able to cause ~1.2% increase of the TSV electrical resistance.

In real situations the probability of growth of the spherical void inside TSV is not high. Void will grow rather along the interface, which is a path of fast migration of vacancies. For an accurate description of void growth kinetics the new equations describing void surface drift should be used. A comprehensive review of these processes is presented in Ref.[19].

As it was mentioned above, the void growth is governed by the incoming flux of vacancies, i.e. the void is considered as a sink for vacancies. The flux of vacancies incoming into the void is proportional to the gradient of the hydrostatic stress. Simultaneously, the vacancy flux along the void surface is proportional to the gradient of surface curvature K. These fluxes can be described by the following formulas [20]:

$$J_{incom} = -\frac{M_s}{h_s} \nabla \sigma_{Hyd}$$
$$J_{surf} = \frac{D_s}{kT} \frac{\partial}{\partial s} \Omega g K \tag{16}$$

Here M_s, h_s, D_s, and g are the probability of void surface penetrating by a vacancy, thickness of the surface layer, surface diffusivity, and surface energy correspondingly. The resulting normal velocity of the void surface is

$$v_n = \Omega \left(J_{incom} + \frac{\partial J_{surf}}{\partial s} \right) \tag{17}$$

Stress evolution equations (1)-(12), coupled with the equations (16) and (17) for void surface motion, were used for simulating the void growth during stress relaxation from 400 MPa to 0.7 MPa (Fig.8,a).

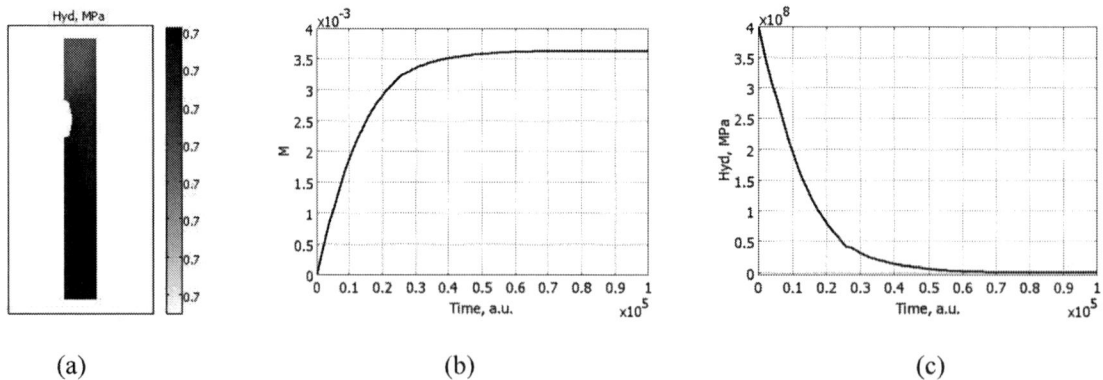

(a) (b) (c)

FIGURE 8. Voiding inside TSV during SM: (a) flaw on the surface transformed to a void ; (b) increase of the average concentration of plated atoms in TSV; (c) decrease of the initial tensile hydrostatic stress.

As it is shown in Fig.8,b, the average concentration of plated atoms grows during the void growth and achieves ~3.6·10⁻³. This process represents the mechanism of initial tensile stress relaxation. As the void traps vacancies, which migrate to the surface due to stress gradients, the concentration of vacancies in TSV should decrease. But the vacancies must be in equilibrium with stress, so new vacancies and corresponding plated atoms are generated. New vacancies keep to be trapped by the growing void, while the plated atoms are immobile; they create compressive stress which compensates the initial tensile stress. So, self-consistent processes of plated atoms concentration increase and stress decrease (Fig.8,c) accompany stress induced voiding in the metal.

EM INDUCED VOIDING IN TSV LANDING PAD

Another element of 3D IC circuit, which should be analyzed regarding the EM behavior, is the TSV landing pad. A non-trivial effect of EM-induced voiding, which was happened inside the landing pad connected to M1 BRDL layer, on the total TSV resistance was observed in [21]. Void was nucleated somewhere near the center of the TSV/pad contact interface (Fig.9,a), then it grew and spanned the whole TSV/pad contact area. Unlike the interconnect wires, where voiding is accompanied by an initial jump in resistance followed by its linear increase with time, in the case of voiding in the pad area located under TSV there is no notable resistance jump after the latency period, and the resistance increases logarithmically with time.

In order to simulate this voiding process, we have modeled the system TSV/pad/M1-wire (Fig.9,a). The considered above stress evolution process inside the TSV has a negligible impact on EM-induced stress build-up in the landing pad. Therefore, the later process can be studied independently. Note that due to the large cross-section of the TSV, which results in the small current densities, EM-induced failure of TSV is significantly less probable than the failure of the M1 wire and the landing pad. In out simulations the values of current density were 9.2mA/um² for M1 layer, and 0.95mA/um² for the TSV.

Figure 9,b demonstrates the simulated highly non-uniform distribution of the electrical current density in the pad area just under TSV. Here, the zero point of the X-axis corresponds to the left edge of the TSV. Current crowding, taking place at the TSV right corner (site A) provides its rapid increase in the direction from the TSV center (site B) toward its right edge (site A). Meanwhile, the curves demonstrating the time evolution of the stress, show that the peak value of the stress is shifting gradually from the TSV edge (site A) to the point near the TSV center (site B).

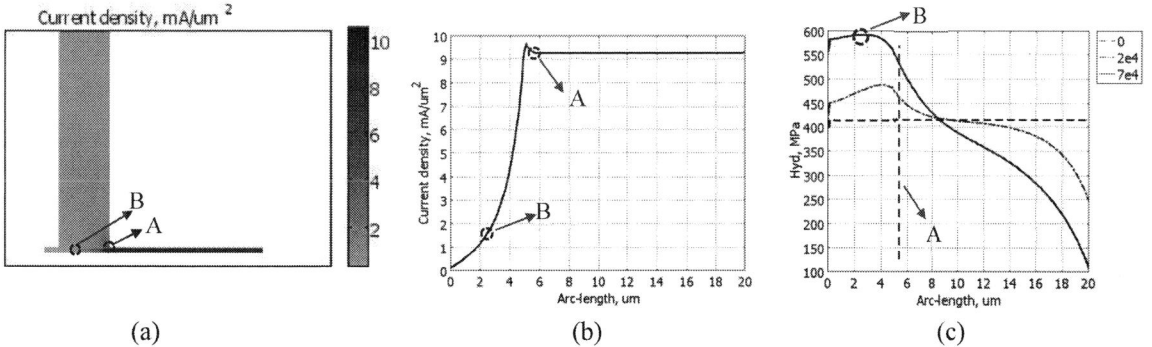

FIGURE 9. The system M1 wire – landing pad – TSV- (a); (b) - distribution of the current density along the wire and pad; (c) – stress evolution along the wire and the pad.

The performed simulation has demonstrated that the critical stress of ~600 MPa, which is presumably required for a void nucleation, for the first time was developed in the pad area located under TSV near the site B, where the current density is very small. The predicted void nucleation location corresponds to one observed in the experiment [21] (Fig.10).

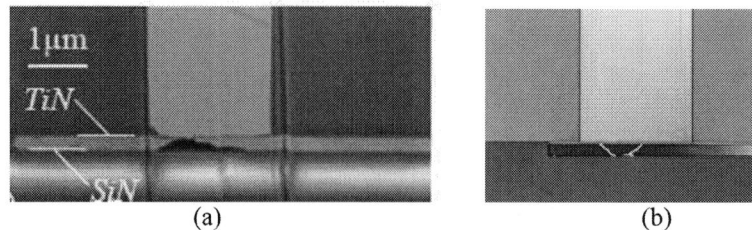

FIGURE 10. Voiding in TSV landing pad: (a) – experiment [18]; (b) – simulations result.

This voiding behavior demonstrates an importance of accounting of the stress gradients-induced fluxes of vacancies in interpretation of the EM-induced failures. As it can be seen in Fig.11, demonstrating the evolution of vacancy concentration, at initial time instances the vacancies are accumulated under the TSV right edge due to the current crowding effect. Stress gradient developed in the pad area underneath the TSV drives the vacancies toward the center of TSV/pad contact area, resulting in the stress increase and void nucleation there.

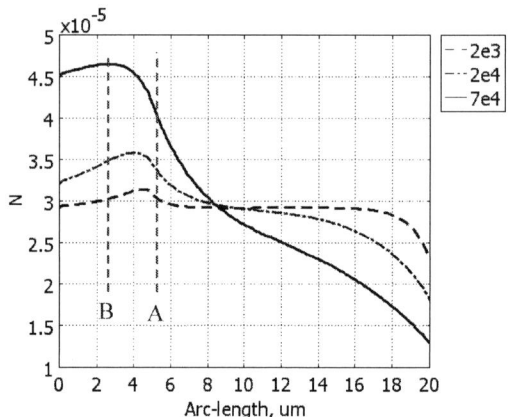

FIGURE 11. Evolution of the concentration of vacancies in the system M1 wire – landing pad.

In order to estimate the change of electrical resistance caused by voiding in the center of the pad/TSV contact area, we need to know the distribution of the electrical potential in the pad. For this purpose the Laplace's equation for the electrical potential V with corresponding boundary conditions should be solved. An approximate distribution along the r – axis can be obtained from 2D analysis (Fig.12(a)):

$$\frac{\partial^2 V}{\partial^2 r} + \frac{\partial^2 V}{\partial^2 z} = 0 \tag{18}$$

Distribution of the potential (before a void nucleation) obtained by FEM simulations is shown in Fig.12(b). The potential varies along the TSV-pad interface. For analytical analysis, the boundary conditions (19) will be used:

$$V(r = 0) = V_0,$$
$$V(r = 2R_{TSV}, z = 0) = 0 \tag{19}$$

where R_{TSV} is the TSV radius. A particular solution of the equation (18) with the boundary conditions (19) is

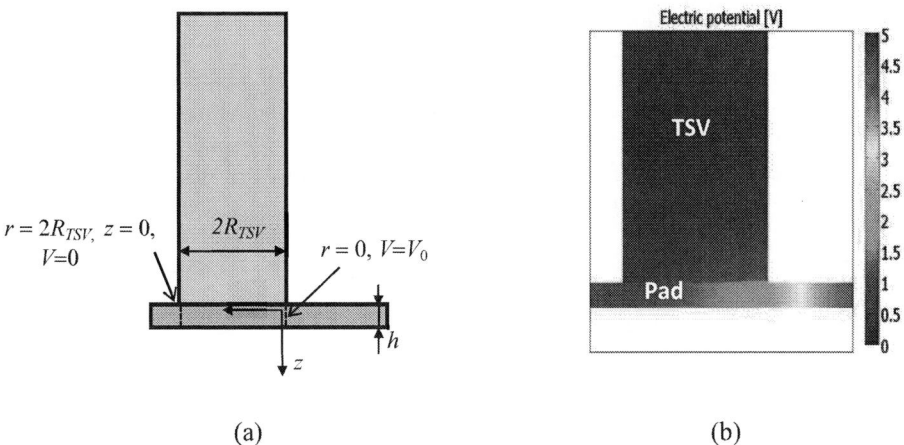

(a)   (b)

FIGURE 12. (a) Schematics of the system TSV – landing pad, used for the analysis of current distribution under the TSV; (b) potential distribution obtained by FEM simulations.

$$V(r,z) = V_0 \left(1 - \frac{2R_{TSV}\, r}{r^2 + z^2} \right) \tag{20}$$

The potential (20) can be used to obtain the distribution of the current density in the pad: $j(r,z) = -\nabla V / \rho$, where ρ is the pad (copper) resistivity.

Now, assuming that origination of cylindrical void in the center of the pad has small impact on the potential distribution (19), we can estimate the average current in the pad under TSV:

$$\bar{I} \sim \frac{1}{R_{TSV}} \int_0^h dz \int_{R_{void}}^{R_{TSV}} |j(r,z)| r\, dr \tag{21}$$

Here h is the pad thickness, which is assumed to be much smaller than TSV diameter. In the case of large void (i.e. $R_{void} > h$) this estimation provides

$$\bar{I} \sim \frac{V_0}{\rho R_{TSV}} \ln \frac{R_{TSV}}{R_{Void}} \tag{22}$$

Assuming that the void radius increases linearly with time $R_{void} \sim t$, the dependency (22) leads to the logarithmic decrease of the current with time, which can be detected in measurements as the increase of the total resistance.

Despite of the roughness of the performed estimation, it demonstrates that the logarithmic change of the resistivity observed in [21] can be observed when the void size is smaller than TSV diameter. When the void spans the whole region under TSV (i.e. $R_{void} = R_{TSV}$), the current can flow only through the liner that covers the pad. According to [21], in this case the resistivity change can still preserve logarithmic character, due to cylindrical symmetry of TSV.

PROSPECTIVE EXPERIMENTAL VALIDATION OF THE SIMULATION MODEL

A design-for-reliability (DFR) type of methodology for managing mechanical stresses in 3D TSV-based chips, stacks and packages is supposed to be based on multi-scale modelling and simulation. Analogously to multi-scale simulation, multi-scale materials parameters are critical as an input for the predictive simulation of the stress distribution across the device layout, and for model validation as well.

One approach to validate the proposed model is based on direct measurement of stress in 3D TSV stacks, preferably the local stress of copper (filled TSV) before and after the anneal process for several positions of the ECD-filled TSV. For Cu grains in the micron and sub-micron range, the strain state of individual copper grains has been reported to be measured using the synchrotron radiation [22]. An alternative approach, not reported in literature so far, could be to measure the strain with nano X-ray diffraction in lab-based nano-XCT tools (the application of nano transmission X-ray microscopy and tomography for imaging has been described in [23]). Modern lab-based nano-XCT tools provide a spatial resolution of 30 nm [24]. For an array of TSVs, conventional or micro X-ray diffraction can be used to measure the depth-dependent strain in Cu TSVs, to determine individual stress values for different parts of the TSV (bottom, middle, top).

The stress measurements in copper (or the surrounding silicon) using X-ray diffraction are advantageous compared to stress measurements in the surrounding silicon using Raman spectroscopy. Micro Raman spectroscopy is a well-established technique to study mechanical stresses in silicon, but the interpretation of the results is not straightforward since all stress components contribute to the frequency shift of the Raman signal, the so-called Raman shift. Often the simple assumption of the "uniaxial stress" is made, and then, the Raman shift depends linearly on the stress. But this assumption is not correct in the case of Si/Cu-TSV systems. Even if the shear stress components could be neglected, which is not the case in the surrounding of the Cu-TSV structures, the Raman shift is related to the difference of tensile and compressive stress components [25].

CONCLUSIONS

Physics-based model of SM and EM kinetics in metal segments has been used for simulation of stress evolution kinetics in TSVs. SM-induced vacancy flow, accompanied by the generation or annihilation of the vacancy – plated atom pairs, results in elimination of the pre-existing stress gradients in TSVs. The process-induced sources of stress gradients, which are the microstructure evolution during high-temperature anneal, and the thermal stress generated by cooling down from the anneal temperature to test temperature, were studied. It was shown that the stress-induced voiding in TSV can be responsible for full relaxation of internal stresses. This process of stress relaxation is caused by accumulation of a large number of plated atoms at GBs and interfaces. EM-induced void nucleation in the TSV landing pad was studied as well. It was demonstrated that the proper modeling of process-induced stresses and stress gradients has allowed us to explain the void nucleation at the site different from the traditional sites where the flux divergence takes place. The latter confirms the importance of the detailed simulations of SM/EM kinetics in complex systems for accurate predictions of a possible failure.

REFERENCES

1. R. Radojcic, M. Nowak, and M. Nakamoto, *AIP Conf. Proc.* **1378**, 5-20 (2011).
2. S. K. Ryu, K. Lu, X. Zhang, J. Im, P. S. Ho, and R. Huang, *IEEE Transactions on Device and Material Reliability* **11**, 35-43 (2011).
3. V. Sukharev, Microelectron. Eng. (2013), http://dx.doi.org/10.1016/j.mee.2013.08.013.
4. V. Sukharev, E. Zschech, *J. Appl. Phys.* **96**, 6337-6343 (2004).
5. V. Sukharev, E. Zschech, and W. D. Nix, *J. Appl. Phys.* **102**, 053505-1 – 053505-14 (2007).
6. P.S. Ho and T. Kwok, *Rep. Prog. Phys.* **52**, 301–348 (1989).
7. C. Herring, *J. Appl. Phys.* **21**, 437-445 (1950).
8. W.W. Mullins, *Metallurgical and Materials Transactions A* **26A**, 1917-1929 (1995).
9. R.J. Gleixner and W.D. Nix, *J. Appl. Phys.* **86**, 1932-1944 (1999).
10. E. Zschech, H.-J. Engelmann, M. A. Meyer, V. Kahlert, A. V. Vairagar, S. G. Mhaisalkar, A. Krishnamoorthy, M. Y. Yan, K. N. Tu, V. Sukharev, *Z. Metallkunde* **96**, 966-971 (2005).
11. M J. Aziz, *Appl. Phys. Lett.*, **70**, 2810-2812 (1997).
12. M. E. Sarychev, Y. V. Zhitnikov, L. Borucki, C.L. Liu, and T.M. Makhviladze, *J. Appl. Phys.* **86**, 3068-3075 (1999).
13. V. Timoshenko and J.N. Goodier, "Theory of elasticity", McGraw-Hill, New York, 1951.
14. V. Sukharev, A. Kteyan, E. Zschech, and W.D. Nix, *IEEE Transactions on Device and Material Reliability* **9**, 87-97 (2009).
15. H. Kadota, R. Kanno, M. Ito, and J. Onuki, *Electrochem. Solid-State Lett.* **14**, D48-D51 (2011).
16. V. Sukharev, A. Kteyan, and E. Zschech, *IEEE Transactions on Device and Material Reliability* **12**, 272-284 (2012).
17. COMSOL, Inc., 8 New England Executive Park, Burlington, MA 01803.

18. Z. Suo, "Reliability of Interconnect Structures" in *Volume 8: Interfacial and Nanoscale Failure*, edited by W. Gerberich, W. Yang, Comprehensive Structural Integrity, Elsevier, Amsterdam, 2003, pp. 265-324.

19. Z. Suo, *Advances in Applied Mechanics* **33**, 194-294 (1997).

20. T. Ogurtani and E.Oren, *J. Appl. Phys.* **102**, 1564-1572 (2001).

21. T. Frank, C. Chappaz, P. Leduc, L. Arnaud, F. Lorut, S. Moreau, A. Thuaire, R. El Farhane, and L. Anghel, IEEE International Reliability Physics Symposium (IPRS), 2011, pp. 3F.4.1 - 3F.4.6.

22. R. Spolenak, N. Tamura, J.R. Patel, *AIP Conference Proceedings* **816**, 288-95 (2006).

23. L. W. Kong, P. Krueger, E. Zschech, A. C. Rudack, S. Arkalgud, A. Diebold, in Stress-Induced Phenomena in Metallization: 11th International Workshop, AIP Proc. 1300, 211 (2010).

24. E. Zschech, W. Yun, G. Schneider, *Appl. Phys.* **A 92**, 423 (2008).

25. I. De Wolf, *AIP Conf. Proc.* **1378**, 138-149 (2011).

TCAD Modeling of Stress Impact on Performance and Reliability in 3D IC Structures

Xiaopeng Xu[a] and Aditya Karmarkar[b]

[a]*Synopsys, Inc. 700 East Middlefield Road, Mountain View, California 94043 USA.*
[b]*Synopsys India Private Limited, Hyderabad, Andhra Pradesh, India.*

Abstract. This paper demonstrates a TCAD based stress modeling approach for analyzing thermal mechanical stress evolution and evaluating stress induced performance and reliability effects in 3D IC structures. A typical 3D IC fabrication and assembly process is examined. It is observed that various TSV and micro-bump process and design parameters need to be optimized in order to minimize the stress impact and fabricate robust 3D structures. TSV copper anisotropy effect is also investigated. Strategies to minimize device performance variation and improve structural reliability are discussed.

Keywords: TSV, bump, 3D IC, TCAD modeling, mechanical stress, carrier mobility, reliability
PACS: 85.40.Qx, 46.70.-p

INTRODUCTION

A typical three-dimensional integrated circuit (3D IC) structure utilizes through-silicon-vias (TSVs) to interconnect silicon dies that are bonded together using micro-bumps [1]. As shown in Figure 1, the stacked structure is a complex assembly of various materials with very different thermal and mechanical properties.

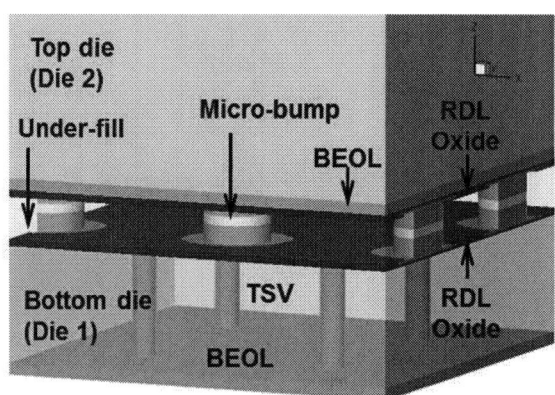

FIGURE 1. 3D integrated structure with TSVs and micro-bumps.

The constituent materials, which include silicon, low k dielectrics, oxide, copper, under-fill, are exposed to multiple thermal ramps during fabrication and stacking. The process steps and thermal ramps introduce thermal mismatch stresses in the stacked structure because of the difference in material coefficients of thermal expansion

(CTEs). For example, the CTE for copper TSV is $17.0 \times 10^{-6}/°C$ while the CTE for silicon is $3.0 \times 10^{-6}/°C$. Large thermal mismatch stresses can be generated after TSV formation. These thermal mismatch stresses add to the material intrinsic stresses that are generated during material formation. All these mechanical stresses evolve along with structural change during fabrication. Die stacking and chip packaging introduce additional mechanical stresses. The final residual stresses modulate carrier mobility in active silicon, affect the structural integrity, and impose reliability concerns [2].

To manage stress in 3D IC structures, a technology computer-aided design (TCAD) modeling methodology needs to be employed [3]. In this paper, a finite element method (FEM) based 3D simulator [4] is used to simulate the entire fabrication and stacking process flow, analyze the stress evolution, and examine the stress induced performance and reliability impact. Sensitivity studies are performed for some important design and technology variables such as TSV diameter and pitch, insulation thickness and stiffness, micro-bump pitch and height, under-fill CTE and modulus.

TCAD SIMULATION OF STRESS EVOLUTION

TCAD Stress Modeling Methodology

Figure 2 shows a typical TCAD simulation flow for stress analysis in a TSV stack. Here, the fabrication process variables and the design layout information in GDSII format are directly used to build 3D structures. The process simulator includes various models to account for material elastic deformation, viscous and plastic flow. It also provides a comprehensive database for material properties. The sequential process simulation is performed together with FEM analysis on the intermediate configurations to capture stress evolution. The numerical solutions for partial differential equations are post processed to examine the impact of stress on the device performance, evaluate the failure probability, and analyze the structural reliability.

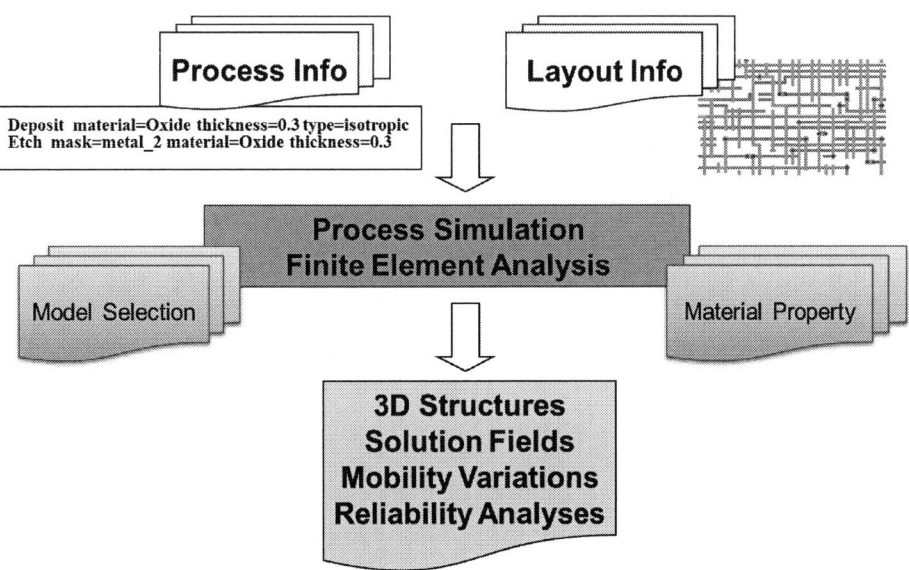

FIGURE 2. TCAD simulation flow for stress analysis in a TSV stack.

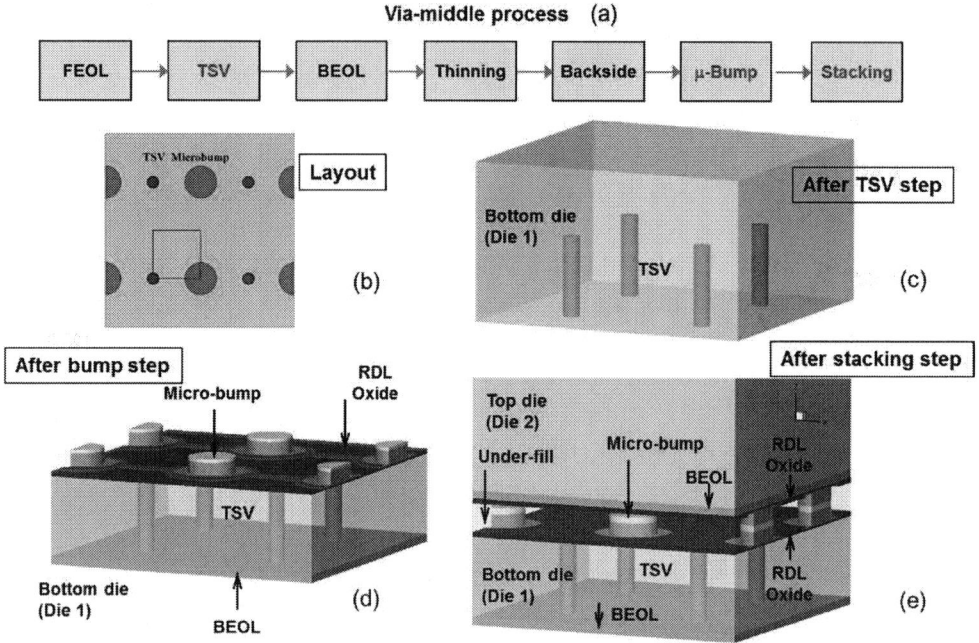

FIGURE 3: Typical TCAD simulation flow for the via-middle process.

Figure 3(a) illustrates a simulation sequence for a 3D IC stack fabricated with the via-middle process. The TSV and micro-bump layout is drawn in Figure 3(b). Figure 3(c) shows the intermediate structure after TSVs are formed inside the bottom die. After back-end-of-line (BEOL) formation, die thinning and backside processing, micro-bumps are formed in the back of the bottom die, as depicted in Figure 3(d). After the completion of the micro-bumps of the top die, two dies are stacked together with under-fills. The simulated 3D structure is shown in Figure 3(e).

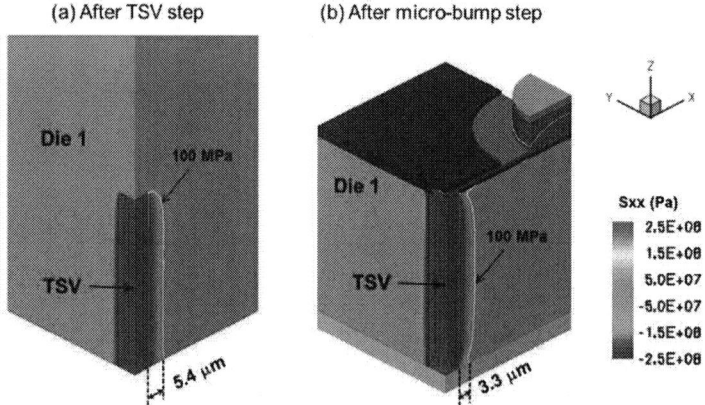

FIGURE 4: Stress evolution with process steps.

Modeling Stress Evolution

The mechanical stress evolution during TSV fabrication and die stacking has to be examined for the stress management with various design parameters in the TSV stack. Figure 4 illustrates the evolution of the normal stress component S_{XX} between two process steps. After TSV formation, the region with Sxx greater than 100MPa extends

130

to 5.4 micron away from TSV. After BEOL formation in the front side of silicon and micro-bump formation in the back side of silicon, the stress distribution is altered with the geometry change. The region with Sxx greater than 100MPa only extends to 3.3 micron away from TSV because of stress evolution with geometry constraint changes.

Similarly, Figure 5 shows the distribution of the hydrostatic stress in the bottom and the top dies before and after stacking. It is evident that the stress magnitudes in both dies have changed during stacking and under-filling process steps due to stress rebalancing under changing geometrical constraints.

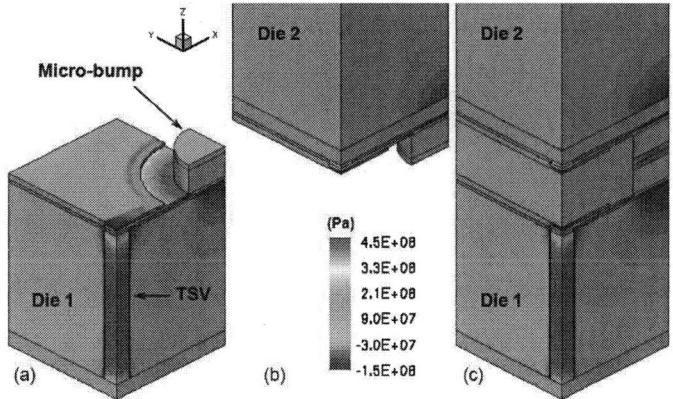

FIGURE 5: Hydrostatic stress before stacking (a) in Die 1, (b) in Die 2, and after stacking (c).

MECHANICAL STRESS IMPACT ON PERFORMANCE

TSV Stress Impact on Performance

The stress induced carrier mobility variation is determined with the pre-stressed state as the reference. The active silicon region has [001]/[110] orientations. The channel direction is along the x-axis with [110] crystal orientation. Figure 6 shows the p-type carrier mobility variation around TSVs on the active silicon surface of Die 1. The largest mobility variation magnitudes occur in the regions surrounding TSVs. The mobility variation magnitude decreases as a function of distance away from the TSV.

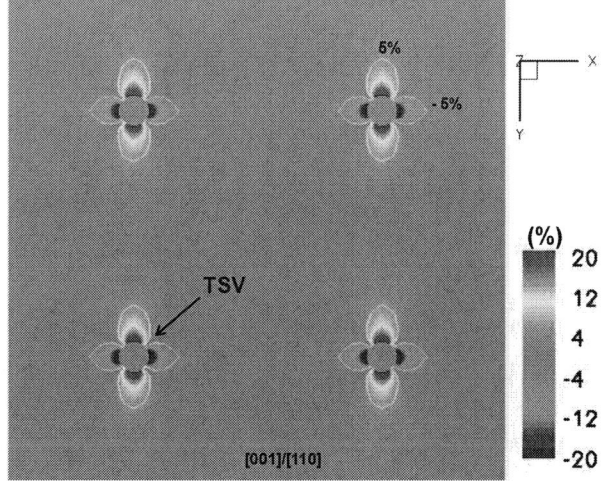

FIGURE 6: p-mobility variation [%] on Die 1 silicon surface.

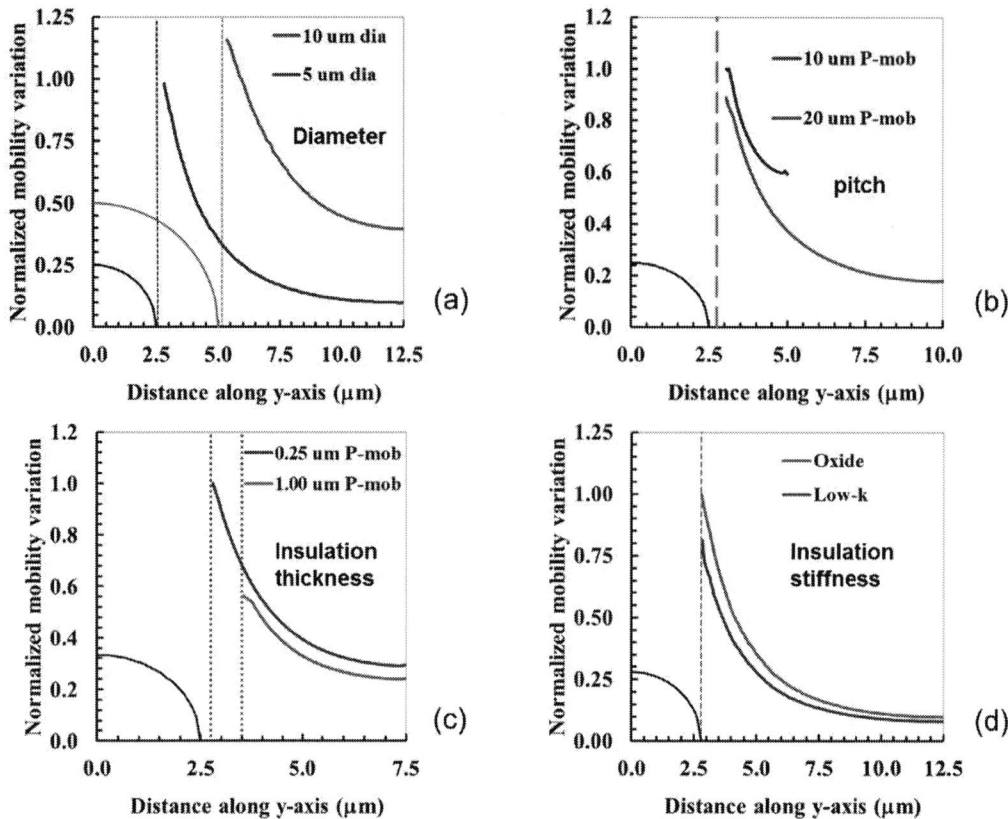

FIGURE 7: TSV material and design parameter impact on the p-type carrier mobility.

In Figure 7, the p-type carrier mobility variations in Die 1 active silicon along the y-axis are examined for various TSV material and design parameters. Figure 7 (a) shows mobility variations for TSV diameters of 5 and 10 µm with a TSV pitch of 25 µm. The variations are normalized to the value observed at the TSV edge for the 5 µm diameter TSV. Larger TSV diameter leads to larger stresses and hence greater the carrier mobility variations. Figure 7 (b) shows the mobility variation for TSV pitches of 10 and 20 µm with a TSV diameter of 5 µm and insulation thickness of 0.5 µm. The variation values are normalized to the value observed at the TSV edge for the pitch of 10 µm. Lower TSV pitch, i.e. closely spaced TSVs, leads to higher carrier mobility variation. Figure 7 (c) shows the mobility variation for TSV insulation thickness of 0.25 and 1 µm with a TSV pitch of 15 µm and diameter of 5 µm. The mobility variation values are normalized to the value observed at the TSV edge for an insulation thickness of 0.25 µm. Thinner TSV insulation results in higher carrier mobility variation. Figure 7 (d) shows the mobility variation for different TSV insulation materials with a TSV diameter of 5 µm, pitch of 25 µm and insulator thickness of 0.3 µm. The mobility variation normalization is with respect to the oxide insulator. Here, oxide insulator shows larger carrier mobility variation than the low-k insulator due to its higher Young's modulus.

In addition to mobility variations, TSV and liner material, process and geometry effects on reliability and insulation performance can also be examined with TCAD modeling and simulation [5].

Micro-bump Stress Impact on Performance

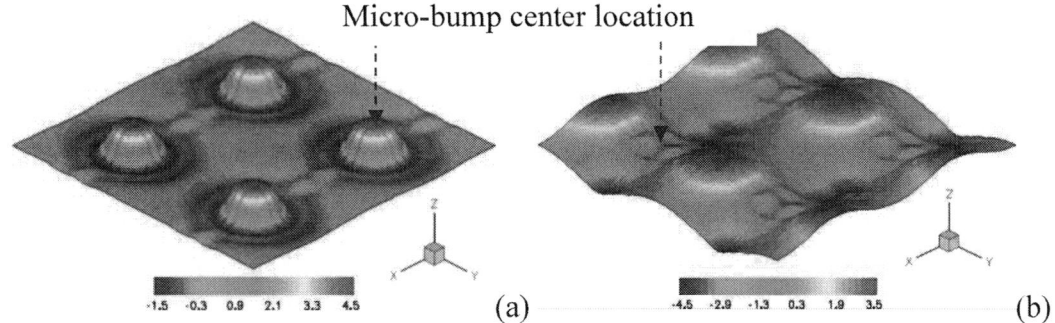

FIGURE 8: Mobility variation [%] on Die 2 active silicon surface. (a) n-type. (b) p-type.

The micro-bump induced stress can introduce carrier mobility variation in Die 2. Figure 8(a) shows the n-type carrier mobility variation on the active surface of Die 2. The largest mobility variation occurs in the regions directly above micro-bumps (see Figure 1). Because of the compression from micro-bumps, the n-type mobility enhancement is observed in these regions. The p-type carrier mobility variation on the active surface of Die 2 is shown in Figure 8(b). It is interesting to note that the largest mobility variation occurs away from the regions right above the micro-bumps. Mobility degradations are bottomed between micro-bumps along the x direction and mobility enhancements are peaked between micro-bumps along the y direction.

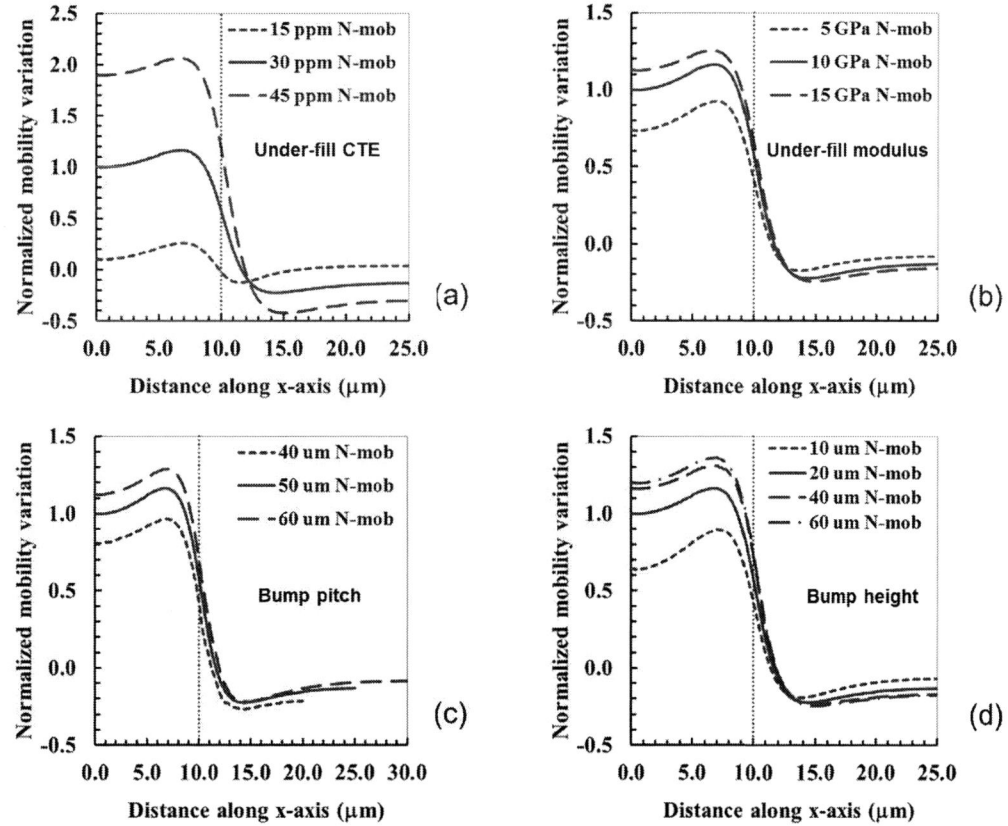

FIGURE 9: Micro-bump material and design parameter impact on the n-type carrier mobility variation.

In Figure 9, the n-type carrier mobility variations in Die 2 active silicon along the x-axis are examined for various under-fill material and micro-bump design parameters. The carrier mobility variation is normalized to the value observed at a point that corresponds to x=0 on the cutline in a 3D model with the bump pitch, radius and height of 50, 10 and 20 μm, respectively, Die 2 height of 50 μm, under-fill Young's modulus of 10 GPa and under-fill CTE of 30 ppm. Figure 9(a) shows mobility variations for various values of under-fill CTEs. The largest under-fill CTE leads to the largest mobility variation. Figure 9(b) plots mobility variations for different values of under-fill Young's modulus. The highest under-fill Young's modulus corresponds to the largest mobility variation. Figure 9(c) depicts mobility variations for different values of micro-bump pitch. The widest micro-bump pitch leads to the largest mobility variation. Figure 9(d) illustrates mobility variation for different values of micro-bump height. The tallest micro-bump corresponds to the largest mobility variation although the mobility variation becomes less sensitive for tall bumps [6].

MECHANICAL STRESS IMPACT ON RELIABILITY

Modeling Stress Impact on Reliability

In a typical 3D IC structure, TSVs and micro-bumps are made of copper. The difference between the thermal expansion coefficients of copper and the surrounding materials, such as insulator, under-fill and silicon, results in large thermal mismatch stresses. These thermal mismatch stresses lead to reliability concerns for the interfaces between copper and surrounding materials [5, 7]. For the same temperature excursion, a copper TSV expands or contracts much more than the surrounding insulator/silicon in the via line direction. This behavior results in the development of large shear stress concentrations at the interfaces especially near silicon surfaces. This shear stress serves as the primary driving force for the formation of cracks and interface delamination. Cracking and interface delamination occurs when the damage driving force exceeds the interface adhesion strength. The damage driving force, i.e. energy release rate, can be examined using the J-integral calculation at the interface of interest [8]. The crack formation and interface delamination can also be analyzed using cohesive zone model [9]. In addition to interface cracking and delamination, micro-bump misalignment induced large shear deformation [10] and TSV process induced effects in electro-migration and stress migration also impose reliability concerns [11].

Effects of Copper Anisotropy

Elastic anisotropy of copper grains in TSV has been observed recently [12]. The observation also reveals the grain size increase from the bottom of TSV to the top. In the extreme case of a few grains or one single grain at the top of the TSV, the anisotropic elastic modulus values along different crystal orientations can be estimated from nano-indentation measurements. Comparing to the copper isotropic Young's modulus value of 112 GPa, the measured anisotropic modulus varies from 69 GPa to 156 GPa along different crystal directions. The copper grain orientation induced elastic modulus anisotropy results in different stress states and distributions around TSVs. The resulting stress state and distribution differences can affect carrier mobility variations in the active silicon regions and crack driving forces in the surrounding materials [13].

Figure 10 shows the n-type carrier mobility variations on Die 1 active silicon surface surrounding one isotropic copper TSV and two anisotropic TSVs. Here, the copper anisotropic properties are defined as three different values of Young's modulus, i.e. E1, E2 and E3, along three orthogonal axes. The Young's moduli for the first anisotropic case, i.e. Anisotropic 1, are 69, 69 and 156 GPa. The Young's moduli for the second anisotropic case, i.e. Anisotropic 2, are 156, 69 and 69 GPa. Comparing to the case with isotropic TSV, the n-type carrier mobility variation around the anisotropic TSV with lower in-plane modulus exhibits reduced variation magnitude; while the n-type carrier mobility variation around the anisotropic TSV with larger in-plane modulus shows increased variation magnitude.

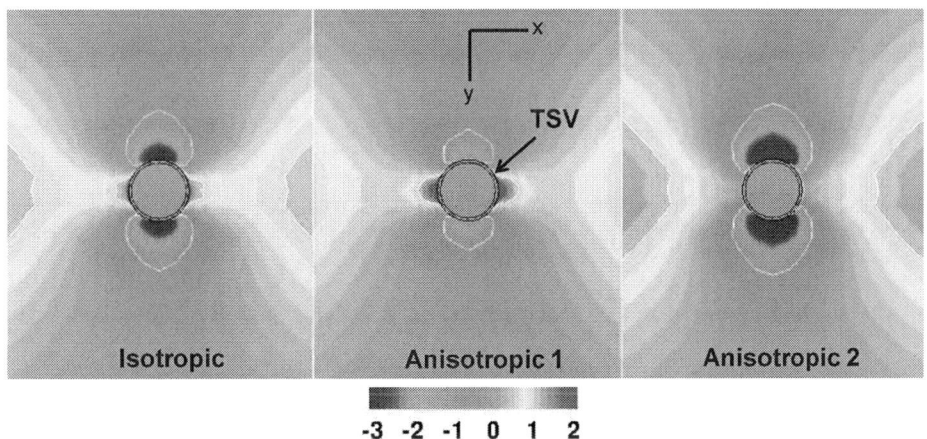

FIGURE 10: Comparison of n-type carrier mobility variations on Die 1 active silicon surface surrounding one isotropic copper TSV and two anisotropic TSVs.

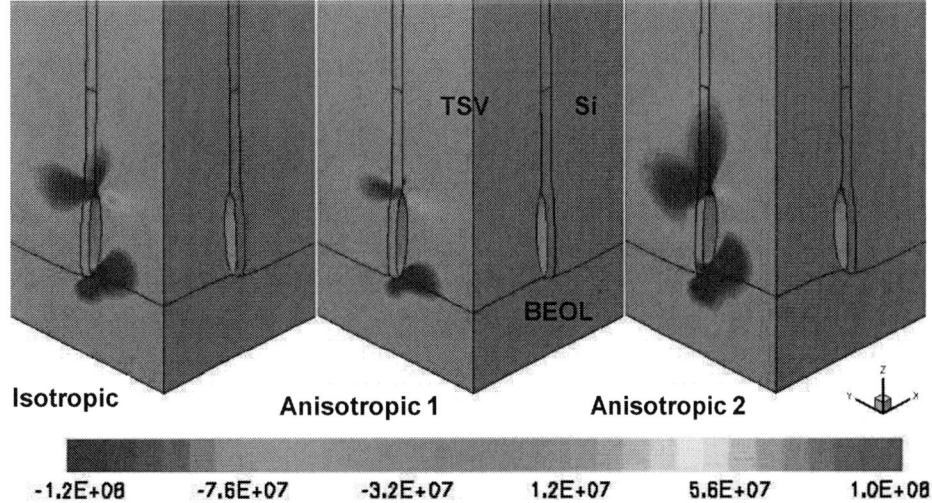

FIGURE 11: Comparison of shear stress component (Syz) [Pa] around a deformed circumferential interfacial crack between TSV and silicon for one isotropic and two anisotropic TSVs.

The TSV copper anisotropy impact on interface crack driving force is illustrated in Figure 11. For the shear stress component Syz around a circumferential interfacial crack, the anisotropic TSV with the largest in-plane modulus yields the highest stress concentration while the anisotropic TSV with the smallest in-plane modulus generates the lowest. It can also be observed that the crack opening shows a mixed mode behavior and it is controlled by both the normal and shear stress components. These results also indicate that the damage driving forces depend on the TSV copper anisotropy and the TSV copper anisotropy significantly affects damage probability at the TSV/insulator/silicon interface.

SUMMARY AND CONCLUSIONS

A typical three-dimensional integrated structure that employs through-silicon vias and micro-bumps is a complex assembly of various sub-structures. These sub-structures consist of different materials with distinct thermal and mechanical properties. During fabrication, stacking and assembly operations, 3D integrated structures are exposed to various thermal excursions. These thermal excursions lead to non-uniform mechanical stress distributions in the 3D structures and these stresses can affect carrier mobility in the active silicon regions and impose structural reliability concerns. In order to design robust 3D systems, it is imperative to accurately evaluate the mechanical stresses and their impact on performance and reliability. Here, an advanced FEM-based 3D TCAD simulator is used to model stress evolution and analyze stress impact on the 3D integration. The simulator accounts for various stress sources such as intrinsic stress, thermal mismatch stress and external loading. The simulator also employs multiple physical models to describe elastic, viscous and plastic material behaviors. Additionally, the simulator directly interfaces with process and layout information, and facilitates technology and design exploration. The simulator generates accurate 3D structures and captures the stress evolution during the entire process sequence. The resulting stress fields are used to analyze the performance and reliability.

In this paper, a typical 3D TSV fabrication and assembly process is examined. It is observed that the mechanical stress distribution in the structure changes with the process steps. It is also observed that the material and process parameters for the TSV and the insulation need to be selected carefully to minimize the device performance impact of mechanical stress in Die 1. Design parameters such as TSV geometry and layout also play a significant role in determining the mechanical stress, and hence must be optimized to lower the stress impact on performance and reliability. The analysis of performance variation due to micro-bump induced stress shows that the micro-bump impact is primarily limited to Die 2 with active surface right above micro-bumps. It is observed that careful optimization of micro-bump parameters leads to minimization of the performance variation due to micro-bump induced stress. Further, TSV anisotropy is observed to affect the performance and reliability of 3D integrated structures. The results indicate that the damage driving forces responsible for interface delamination and cracking depend on TSV anisotropy. These crack driving forces, and hence the failure probability, can be minimized by optimizing TSV process to engineer TSV copper anisotropy.

REFERENCES

1. R. Radojcic, M. Nowak and M. Nakamoto, "TechTuning: Stress management for 3D through-silicon-via stacking technologies", *AIP Conference Proceedings,* vol. 1378, pp. 5-20, 2011.
2. A. P. Karmarkar, X. Xu and V. Moroz, "Performance and reliability analysis of 3D-integration structures employing through silicon via (TSV)", *Proceedings of the 47th Annual IEEE International Reliability Physics Symposium (IRPS)*, pp. 682-687, Apr. 2009.
3. X. Xu and A. P. Karmarkar, "3D TCAD modeling for stress management in through silicon via (TSV) stacks", *AIP Conference Proceedings,* vol. 1378, pp. 53-66, 2011.
4. TCAD Sentaurus Interconnect User Guide, *Synopsys, Inc.*, Sep. 2011.
5. A. P. Karmarkar, X. Xu, S. Ramaswami, J. Dukovic, K. Sapre and A. Bhatnagar, "Material, process and geometry effects on through-silicon via reliability and isolation," *Materials Research Society Symposium Proceedings*, vol. 1249, F09-08, 2010.
6. A. P. Karmarkar and X. Xu, "Microbump impact on reliability and performance in through-silicon via stacks", *Materials Research Society Symposium Proceedings*, vol. 1335, O08-05, 2011.
7. S.-K. Ryu, K.-H. Lu, X. Zhang, J. Im, P. S. Ho, and R. Huang, "Impact of Near-Surface Thermal Stresses on Interfacial Reliability of Through-Silicon-Vias for 3-D Interconnects", *IEEE Trans. on Device and Materials Reliability*, vol. 11, pp. 35-43, 2011.
8. C. F. Shih, B. Moran and T. Nakamura, "Energy release rate along a three-dimensional crack front in a thermally stressed body", *International Journal of Fracture,* vol. 30, no.2, pp. 79 - 102, February 1986.
9. X. Xu and A. Needleman, "Numerical simulations of fast crack growth in brittle solids", *Journal of the Mechanics and Physics of Solids*, vol. 42, no. 9, pp. 1397 - 1434, September 1994.
10. Y.-L. Shen and R.W. Johnson, "Misalignment induced shear deformation in 3D chip stacking: A parametric numerical assessment", to appear in *Microelectronics Reliability*, 2012.
11. V. Sukharev, "TSV process-induced effects in electromigration and stress-migration", to appear in *AIP Conference Proceedings,* 2012.
12. K. B. Yeap, E. Zschech, U. D. Hangen, T. Wyrobek, L. W. Kong, A. Karmarkar, X. Xu, and I. Panchenko, "Elastic anisotropy of Cu and the impact on stress management for 3D IC: Nanoindentation and TCAD simulation study", *Journal of Materials Research*, vol. 27, no.1, pp. 339-348, Jan. 2012.
13. A. P. Karmarkar, X. Xu, K. B. Yeap and E. Zschech, "Copper anisotropy effects in three-dimensional integrated circuits using through-silicon vias", *IEEE Transactions on Device and Materials Reliability*, vol. 12, no. 2, pp. 225 – 232, June 2012.

Design for Reliability of BEoL and 3-D TSV Structures – A Joint Effort of FEA and Innovative Experimental Techniques

Jürgen Auersperg, Dietmar Vogel, Ellen Auerswald,
Sven Rzepka, Bernd Michel

Fraunhofer ENAS, Micro Materials Center, Technologie-Campus 3, D-09126 Chemnitz, Germany

Abstract. Copper-TSVs for 3D-IC-integration generate novel challenges for reliability analysis and prediction, e.g. the need to master multiple failure criteria for combined loading including residual stress, interface delamination, cracking and fatigue issues. So, the thermal expansion mismatch between copper and silicon leads to a stress situation in silicon surrounding the TSVs which is influencing the electron mobility and as a result the transient behavior of transistors. Furthermore, pumping and protrusion of copper is a challenge for Back-end of Line (BEoL) layers of advanced CMOS technologies already during manufacturing. These effects depend highly on the temperature dependent elastic-plastic behavior of the TSV-copper and the residual stresses determined by the electro deposition chemistry and annealing conditions. That's why the authors pushed combined simulative/experimental approaches to extract the Young's-modulus, initial yield stress and hardening coefficients in copper-TSVs from nanoindentation experiments, as well as the temperature dependent initial yield stress and hardening coefficients from bow measurements due to electroplated thin copper films on silicon under thermal cycling conditions. A FIB trench technique combined with digital image correlation is furthermore used to capture the residual stress state near the surface of TSVs. The extracted properties are discussed and used accordingly to investigate the pumping and protrusion of copper-TSVs during thermal cycling. Moreover, the cracking and delamination risks caused by the elevated temperature variation during BEoL ILD deposition are investigated with the help of fracture mechanics approaches.

Keywords: 3-integration, copper-TSV, nonlinear FEA, BEoL, interface fracture mechanics
PACS: 07.05.Pj; 46.50.+a; 46.80.+j; 81.15.Pq; 81.40.Lm; 81.70.Bt; 85.40.-e; 85.40.Ls; 85.40.-e

INTRODUCTION

Copper is the most preferred interconnect material used in **t**hrough **s**ilicon **v**ias (TSV) as it provides low electrical resistivity, high electro migration resistance and high current-carrying capacity, which are best preconditions for use in 3D-IC-integration. On the other hand, it causes novel challenges for reliability analysis and prediction, e.g. the need to master multiple failure criteria for combined loading together with residual stress, interface delamination, cracking and fatigue issues. The prime cause is the thermal expansion mismatch of copper to silicon. So, copper-TSV causes varying stresses in the silicon surface surrounding the TSVs, which is influencing the electron mobility and as a result the transient behavior of transistors. Furthermore, pumping and protrusion of copper is a challenge for Back-end of Line

(BEoL) layers of advanced CMOS technologies already during manufacturing. These effects depend highly on the temperature dependent elastic-plastic behavior of TSV-copper and the residual stresses determined by the electro deposition chemistry and annealing conditions.

There are various published investigations on cracking risks in the surrounding of copper-TSVs and also on the influence copper-TSVs on the transient behavior of transistors [1]. All these investigations base on quite different linear elastic or elastic plastic copper properties and stress free assumptions. An inspection of the material properties from literature shows a wide spreading of data. It indicates clearly the importance of determining appropriate properties for the chemistry, the filling and annealing conditions actually used during TSV manufacturing. Investigations to extract the Young's-modulus, initial yield stress and hardening coefficients from nanoindentation experiments on copper TSVs are necessary, as well as the measurement of the temperature dependent initial yield stress and hardening coefficients of electroplated thin copper films.

These properties give a qualified base for the investigation of the pumping and protrusion phenomena of copper-TSVs during thermal cycling. Furthermore, they are preconditions to investigate the cracking and delamination risks caused by the elevated temperature variation during BEoL ILD deposition with the help of fracture mechanics approaches.

Currently, even without a complete comparison with silicon stress measurements in the vicinity of copper-TSVs by nanoRaman spectroscopy or related methods, the stringent continuous concept via combined simulative/experimental parameter identification approaches gives a confidence-inspiring base for further optimizations concerning copper-TSVs for 3D-IC-integration.

MECHANICAL BEHAVIOR OF TSV-COPPER

Copper as used within TSVs is usually deposited via electroplating technology. Several supplier specific chemistries and annealing conditions lead to wide strewing thermo-mechanical properties. This fact, together with the different samples and loading conditions selected to measure them, explains the amount of publications and the resulting variety of measured data. Also the assumed thermo-mechanical constitutive behavior varies between pure elastic to elastic plastic, isotropic or/and kinematic hardening material behavior [1-4]. It has to be noted, that numerous publications don't take into account the temperature dependence of copper material properties as well as their size dependence (Hall-Petch effect or strain-gradient plasticity assumptions). Especially investigations on keep-out zones for actives surrounding Cu-TSVs often utilize elastic isotropic behavior of copper [5].

But, measurements performed by IMEC (see [6]) and described in detail by Okoro et al. [7] suggest that electroplated Cu behaves as particular inelastic, kinematic hardening, elastic-plastic material. These measurements based on wafer curvature measurements on thin film copper, deposited on silicon dies during repeated thermal cycles between room temperature (RT) and 420 °C. They showed that the stress/deformation behavior in the first heating up phase to 420 °C is quite different to all further heating/cooling processes. This indicates morphological changes in the Cu-TSV, see Fig. 1.

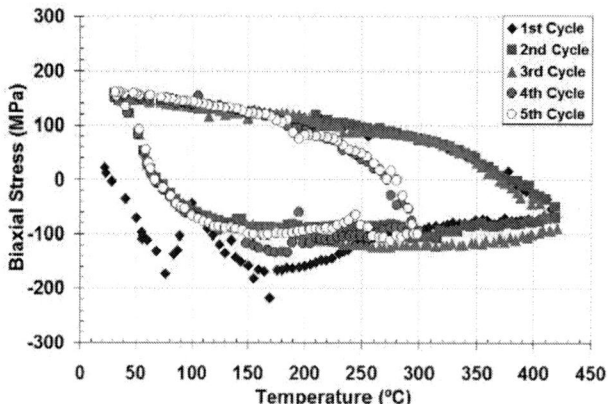

FIGURE 1. Stress evolution as a function of temperature during thermal cycling of electroplated Cu-film (IMEC – see [4])

With the intention to establish a well defined data base for the thermo-mechanical behavior of a dedicated TSV-Cu, the authors started nanoindentation experiments to extract the Young's modulus at the upper surface of TSVs. It was found to be in the range between E = 105 GPa and E = 140 GPa. As well known from literature, it reduces with increasing temperature to 95 GPa and 70 GPa at 350 °C and higher temperatures. Experiments, alike those in [7] and with results like in Fig. 1, where performed and supported by numerical simulations, which are equivalent in geometry, material and thermal loading conditions. In our case the samples are strips of silicon wafers with a 5 µm copper film on top. The bending of the compound was measured prior to and during thermal cycling and biaxial stresses were calculated via a modified Stoney's formula [8].

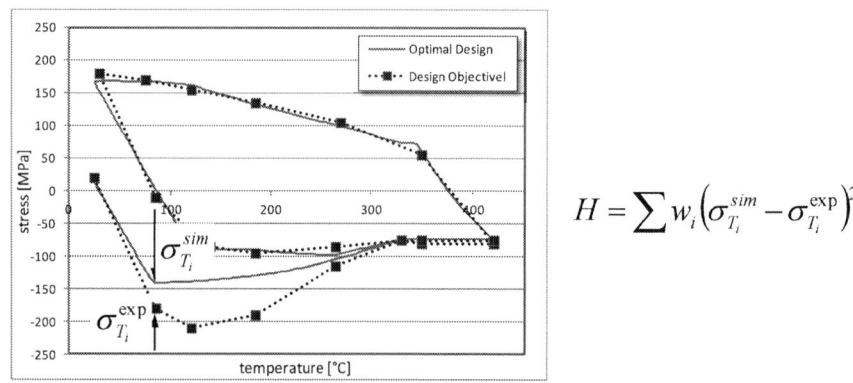

$$H = \sum w_i \left(\sigma_{T_i}^{sim} - \sigma_{T_i}^{exp} \right)^2$$

FIGURE 2. Optimization approach to fit FEA- and experimental results for thin Cu-film on silicon thermal cycling experiments

Appropriate finite element analyses deliver the biaxial copper-stress vs. temperature curves as shown in Fig 2. Assuming nonlinear, elastic plastic, kinematic hardening behavior of copper an optimization approach was established to obtain the material properties fitting to the copper film under investigation. This approach utilizes the finite element model as testing vehicle to reach the best approximation between FEA and experimental results. Therefore, the initial yield stresses and the hardening coefficient at several discrete temperatures act as optimization parameters (variables). The objective function H is the sum of squares of the errors between measured and calculated film-stress at every temperature taken into account - see Fig. 2. Natural inspired global optimization and natural inspired local refinement and gradient based optimization algorithms reduced the value of H by playing with σ_y (T_j) and E_p (T_j), where T_j are the discrete temperatures defining the plastic straining behavior of copper.

As results, we found the initial yield stress as a function of temperature in the range of σ_y = 200 MPa and σ_y = 100 MPa decreasing towards higher temperatures. Secondly, the hardening modulus was found between E_p = 3000 MPa and E_p = 500 MPa - also decreasing towards higher temperatures. The determined material model describes very well the hysteresis loop and also the remaining residual stresses at room temperature (RT) after the 1st heating up. However, it was necessary to reduce the weighting factors w_i for all points that describe the 1st heating up to 425 °C, in order to ensure appropriate results for the hysteresis loop. This right different behavior of the Cu-film during 1st heating up is surely due to changes in the copper morphology – grain size, grain orientation – an effect of annealing. It suggests for several samples that those processes start already at about 350 °C. Thermal handling subsequent to a 1st heating up to 425 °C is not significantly changing the residual stress at RT – a tensile stress of about 160 MPa is the result.

EBSD STRUCTURAL ANALYSES

Just to make the structural changes in TSV-copper more clear, experimental investigation by means of EBSD (Electron Backscattering Diffraction) has been realized at cross sections of several TSVs.

  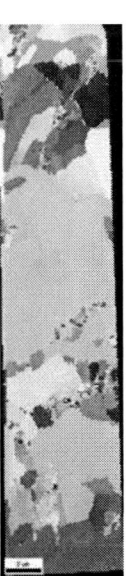

FIGURE 3. Inverse pole figures of TSV-copper

Undisposed   1. Heating   5. Heating

Grain size maps taken before and after thermal cycling show an increase of the copper grain sizes after the 1st and the 5th thermal cycle (cycling between RT and 350 °C) – see inverse pole figures in Fig. 3. Also, a change in crystal orientation is shown already after the 1st heating cycle – see Kernel Average Misorientation (KAM) in Fig. 4. So from modeling point of view, the crystal orientation changes during those thermal cycles suggest a change from more or less isotropic to anisotropic behavior of TSV-copper during thermal handling. This is a conclusion made from the volume fraction reduction of the kernel average misorientation (KAM) from 0° to 1° after the 1st heating and an increase of the volume fraction of KAM from 1° to 2° during thermal cycling. All these results point to recrystallization activities during thermal handling, which could be one of the leading mechanisms driving the highly discussed TSV-protrusion effect.

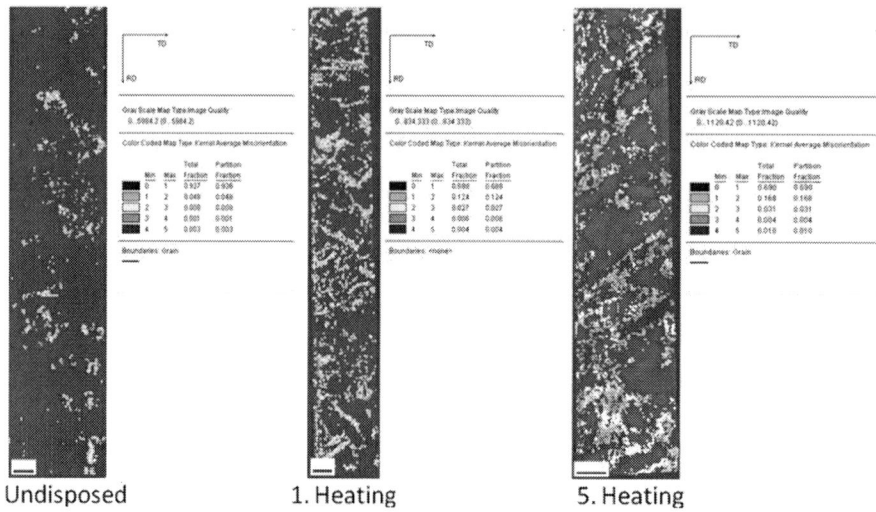

Undisposed   1. Heating   5. Heating

FIGURE 4. Kernel Average Misorientation of TSV

RESIDUAL STRESS IDENTIFICATION

As numerical investigations pointed out, knowledge of the initial stresses are essential for properly interpreting the stress/strain fields and the motion of copper in TSVs against the surrounding silicon. The authors developed and used a new technique based on stress release by FIB milling. Digital image correlation (DIC) algorithms are utilized to determine stress release deformations from SEM micrographs, captured before and after ion milling. As a result residual stresses can be computed [13-14].

First attempts were made by the authors, milling tiny trenches into the surface of TSVs of interest. It is the intention of the method, to use trench opening/closing as a measure of released stresses. Therefore, finite element analysis was applied in combination with displacement field measurements around the milled trench. Both displacement fields from FEA and DIC application, respectively, have to be compared by some kind of least square fit in order to determine the stress values describing the experimental data in a best way. Material properties stem from investigations previously exposed. So, the initial stresses after copper deposition represent the principal parameter of the FE-model with direct influence on sequential simulation results.

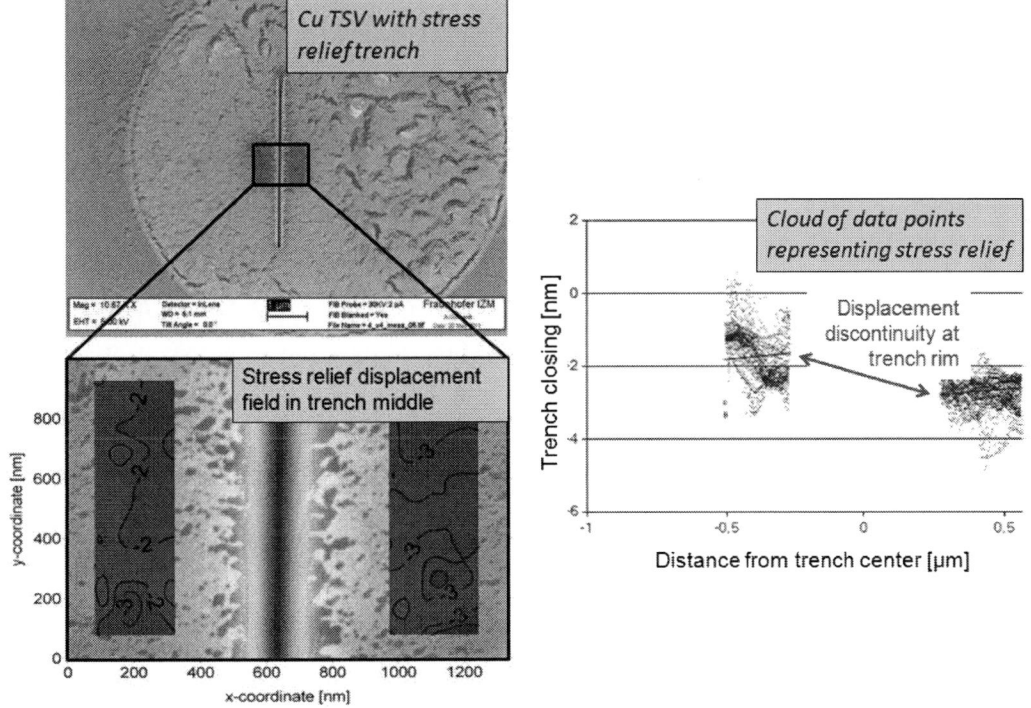

FIGURE 5. Trench in TSV surface and trench opening as a result of stress relief (left and the Cloud of data points representing stress relief

It is to be noted, that additional "stress measurements" by means of Raman spectroscopy in the surrounding silicon surface near TSVs have been carried out to support the modeling assumptions, as well. For subsequent numerical studies several assumptions regarding initial stresses were utilized.

STUDYING TSV PROTRUSION

The thermal expansion mismatch between copper and silicon well-known is the major challenge for using copper-TSV in 3D-IC integration. It can lead to delamination between TSV and the surrounding liner/silicon. In addition, the vertical pumping of the TSV-surface during thermal cycling can lead to damage or/and delamination of already deposited layers of the front-side BEoL-stack – see Ryu et al. [10]. The material model previously described is used to simulate thermal cycling between RT and 100 °C to study the physics behind the protrusion effect regarding several modeling assumptions.

The loading history starts with a heating up from RT to 425 °C followed by cooling down back to RT and 100 thermal cycles from -64 °C to 150 °C. This is to fulfill the conditions during material property fitting and should match the annealing effect during 1st heating up. On the other hand and to be checked furthermore, there is a tensile residual stress of about 160 MPa in the TSV at RT. Model variations are:

1. Use of the elastic plastic, kinematic hardening material model,
2. Use of a ductile damage model in addition to assumption 1,
3. Add CZM- (cohesive zone model) elements into the interface TSV-Cu/Liner in addition to assumption 1.

Fig. 6 depicts the TSV-pumping and -protrusion vs. thermal cycling tests (TCT) after the 1st heating up to 425 °C and cooling down to RT, as mentioned above. Whilst only the elastic plastic, kinematic hardening material model is acting here, the results for the two other cases look similar. This means, the pumping amplitude is slightly higher than the protrusion (the changing vertical position of the TSV-surface against the Si-surface at RT per thermal cycle). First two thermal cycles (after the 1st heating up to 425 °C and cooling down to RT) fulfill the major motion with a remaining positive protrusion, which depends strongly on the stress-free state assumptions.

We set hypothetically the TSV-protrusion as zero at that point assuming grinding/CMP is used to reach a flat surface. As a result, this adjusted protrusion amplitude is less than 100 nm and negative at all times. The reason is surely the tensile residual stress in the TSV at RT, as introduced prior to TCT. This is a result of the first heating up to 425 °C and cooling down to RT. In addition, the plastic strains accumulate, being strongly localized near the upper part of the interface of the TSV to the liner – see Fig. 7.

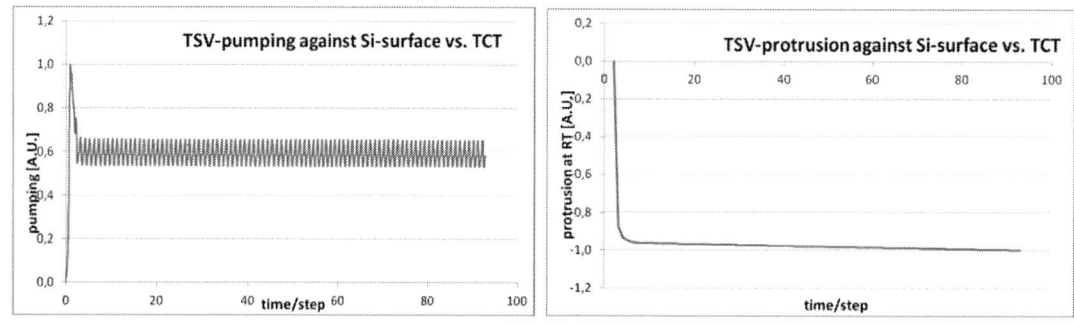

FIGURE 6. TSV-pumping (left) and –protrusion (right) assuming elastic plastic, kinematic hardening behaving copper

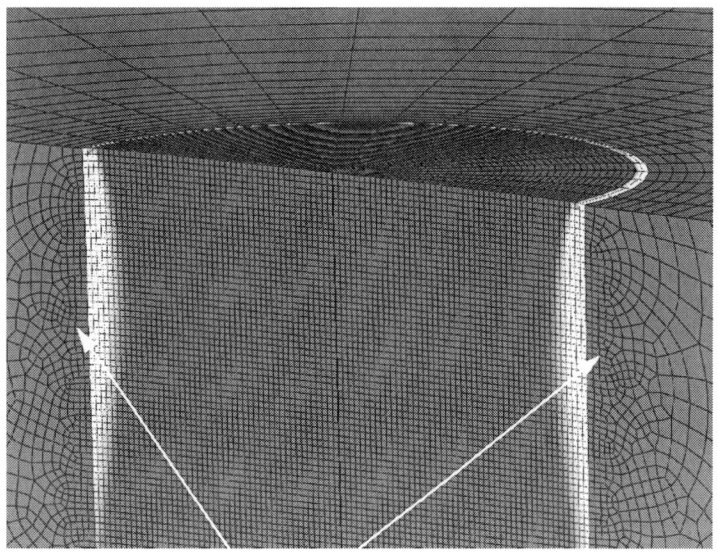

FIGURE 7. Localized plastic strain accumulation in copper-TSV

TSV PROTRUSION AND BEOL DELAMINATION

The thermal expansion mismatch between the main competitors copper and silicon together is responsible for protrusion, considering the kinematic hardening elastic plastic behavior, damaging of TSV-copper, delamination and – possibly – changes in grain sizes and crystal orientation. So, questions regarding the influence of the expected protrusion of copper on the delamination and cracking risk of BEoL-structures deposited on top of TSVs remain unsolved. For this subject, fracture mechanics concepts seem to provide good measures for process and design evaluation and optimization – unfortunately with some drawbacks:

1. Delamination between copper and barrier layers is possible, but the investigation by integral concepts or virtual crack closure technique (VCCT) is handicapped by the huge extension of plastic deformations in copper.
2. Protrusion as a result of high temperatures during BEoL-processing is highly dependent on the area of those potential delaminations and the initial stress state in copper after its insertion.
3. Crack flanks of potential delamination between several structures of MOL and BEoL close up during several steps of processing – the necessary contact handling conflicts with fracture mechanical concepts.
4. J-integral or VCCT seem to be unusable for the frequently alternating stresses (direction of traction vector alternates) during several steps of processing.
5. Mode separations fail or come to nothing, as expected.

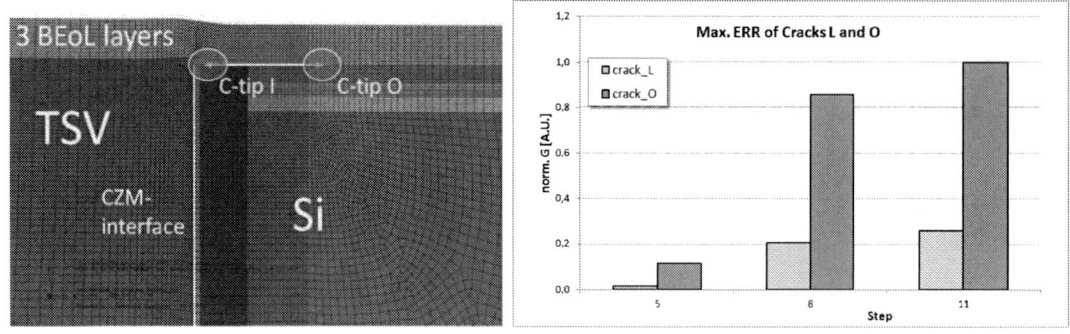

FIGURE 8. Initial cracks assumed for delamination initiation/propagation investigations (left), Max. energy release rate (ERR) of crack tips L and O vs. BEoL deposition
(right, 1st ... step 5, 2nd ... step 8, 3rd ... step 11)

The use of fracture mechanics concepts - as started in [11] and explained in [12] with regards to the concepts comparability - for investigations of TSV delamination shows, that an assumed initial crack underneath the 1st BEoL-layer is more critical to initiate outwards and the deposition of the 1st 1-2 layers seems more critical than subsequent layers.

But, even if the fracture concepts deliver comprehensible results, their validity is not proven up to now and the models themselves have also to be improved. This is especially due to an appropriate reflection of the grinding/CMP before BEoL-deposition. Moreover, the residual stress state at this step of processing is not checked experimentally. For this purpose a FIB trench technique combined with digital image correlation will be used in the near future to capture the residual stress state on the surface of TSVs.

CONCLUSION

Copper-TSVs for 3D-IC-integration generate novel challenges for reliability analysis and prediction, i.e. the need to master multiple failure criteria for combined loading including residual stress, interface delamination, cracking and fatigue issues. Furthermore, pumping and protrusion of copper is a challenge for BEoL-layers of advanced CMOS technologies already during manufacturing. These effects depend highly on the temperature dependent elastic-plastic behavior of the TSV copper and the residual stresses determined by the electro deposition chemistry and annealing conditions.

The authors pushed a combined simulative/experimental approach to extract the Young's-modulus, initial yield stress and hardening coefficients from nanoindentation experiments on copper TSVs and the temperature dependent initial yield stress and hardening coefficients from bow measurements of electroplated thin copper films on silicon under thermal cycling conditions. The extracted properties were discussed and used accordingly to investigate the pumping and protrusion of copper-TSVs during thermal cycling. Furthermore, the cracking and delamination risks caused by the elevated temperature variation during BEoL ILD deposition are investigated with the help of fracture mechanics approaches, in particular.

ACKNOWLEDGMENTS

The authors would like to thank Saskia Huber from Fraunhofer IZM, Berlin and Raúl Mroßko from AMIC GmbH, Berlin for their contribution to the nanoIndentation tests. The authors would also like to thank Kashi Vishwanath Machani from GLOBALFOUNDRIES Dresden Module One LLC & Co. KG for the fruitful discussions and guiding through the technological challenges.

REFERENCES

1. Jiang, TF et al., Applied Physics Letters, Volume **103**, 211906 (2013)
2. N. Ranganathan, K. Prasad, N. Balasubramanian and K. L. Pey, in Journal of Micromechanics and Microengineering 18 (July 2008)
3. M. Wolf, B. Wunderle, N. Jürgensen, G. Engelmann, O. Ehrmann, B. Michel, Th. Dretschkow, A. Uhlig, H. Reichl, in Proceedings of the *58th Electronic, Components & Technology Conference*, Orlando, FL, 2008, pp. 563–570
4. J. Lau, *The McGraw-Hill Comp.* © 2011, pp. 453-504
5. C. Okoro, Y. Yang, B. Vandevelde, B. Swinnen, D. Vandepitte, B. Verlinden, I. De Wolf, in Proceedings of the *International Interconnect Technology Conference, Burlingame, CA, USA, 2008*
6. J.-S. Yang, K. Athikulwongse, Y.-J. Lee, S. K. Lim, D. Z. Pan, in Proceedings of the *47th Annual Design Automation Conference,* 2010, Anaheim, Ca., USA, pp. 803-806
7. Imec, Scientific report 2010, http://www.imec.be/ScientificReport/SR2010/2010/1159173.html
8. Okoro, Ch., Labie, R., Vanstreels, K., Franquet, A., Gonzales, M., Vandevelde, B., Beyne, E., Vandepitte, D., Verlinden, B., in *Journal of Material Science (2011)* 46:3868–3882
9. G. Stoney, in Proceedings *Royal Soc.* London, A82 (1909) 172
10. S.-K. Ryu, K.-H. Lu, X. Zhang, J.-Hi Im, Ho, P.S.; R. Huang, in *IEEE Trans. On Device and Materials Reliability*, Vol. 11, No. 1, March 2011, pp. 35-43
11. J. Auersperg, R. Dudek, B. Michel, in Proceedings of the *12th Electronics Packaging Technology Conference* Singapore 2010
12. J. Auersperg, R. Dudek, J. Oswald, B. Michel, in Proceedings of the *12th Int. Conf. on Thermal, Mechanical and Multiphysics Simulation and Experiments in Micro-Electronics and Micro-Systems,* Linz, Austria, 2011
13. N. Sabaté, D. Vogel, A. Gollhardt, J. Keller, B. Michel, C. Cané, I. Gràcia, J. R. Morante, in *Appl. Physics Letters* 88, 071910, 2006
14. D. Vogel, A. Gollhardt, N. Sabate, J. Keller, B. Michel, H. Reichl, in Proceedings of the 57th *Electronic Components and Technology Conference,* Reno, USA, 2007, pp. 1490-1497

Thermomechanical Characterization and Modeling for TSV Structures

Tengfei Jiang[a], Suk-Kyu Ryu[b], Qiu Zhao[a], Jay Im[a], Paul S. Ho[a], and Rui Huang[b]

[a]*Microelectronics Research Center, University of Texas, Austin, TX 78712*
[b]*Department of Aerospace Engineering and Engineering Mechanics, University of Texas, Austin, TX 78712*

Abstract. Continual scaling of devices and on-chip wiring has brought significant challenges for materials and processes beyond the 32-nm technology node in microelectronics. Recently, three-dimensional (3-D) integration with through-silicon vias (TSVs) has emerged as an effective solution to meet the future technology requirements. Among others, thermo-mechanical reliability is a key concern for the development of TSV structures used in die stacking as 3-D interconnects. This paper presents experimental measurements of the thermal stresses in TSV structures and analyses of interfacial reliability. The micro-Raman measurements were made to characterize the local distribution of the near-surface stresses in Si around TSVs. On the other hand, the precision wafer curvature technique was employed to measure the average stress and deformation in the TSV structures subject to thermal cycling. To understand the elastic and plastic behavior of TSVs, the microstructural evolution of the Cu vias was analyzed using focused ion beam (FIB) and electron backscattering diffraction (EBSD) techniques. Furthermore, the impact of thermal stresses on interfacial reliability of TSV structures was investigated by a shear-lag cohesive zone model that predicts the critical temperatures and critical via diameters.

Keywords: through-silicon via; thermal stress; via extrusion; interfacial reliability.

INTRODUCTION

Through-silicon-via (TSV) is a key element for 3-D integration in providing vertical interconnects for chip-stacking structures. Copper (Cu) is widely used as the via filling material because it is compatible with both the front-end of line (FEOL) and back-end of line (BEOL) processes. However, the mismatch in the coefficients of thermal expansion (CTEs) between Cu and Si is relatively large, which is responsible for the development of thermal stresses in the TSV structures. The thermal stresses can arise during fabrication, testing and service of the TSVs, leading to various reliability issues, such as crack growth, via extrusion, and degradation of device performance [1-4]. Therefore, it is important to experimentally characterize the thermal stresses and understand their impact on TSV reliability for development of 3-D interconnects.

This paper is organized in two parts. The first part presents the experimental methods for stress characterization. A precision wafer curvature technique has recently been applied to measure the thermal stresses of TSV structures during thermal cycling [5]. As a global measurement, the curvature change provides a measure of the average thermal stresses. The behavior can be correlated to the evolution of the Cu microstructure in the via when subjected to thermal processing, which was analyzed by focused ion beam (FIB) and electron backscatter diffraction (EBSD) techniques [6]. Based on the results from the microstructure analysis, the mechanisms underlying the linear and nonlinear temperature-curvature behavior of the TSV specimen are discussed. The local stress distribution near the Si surface around the Cu vias is important on device performance and interfacial reliability [7, 8]. This was measured by micro-Raman spectroscopy [9]. The stresses measured by Raman spectroscopy can in turn be correlated to that observed by the wafer curvature method through finite element analysis (FEA), taking into account the reference temperature, which was also based on the wafer curvature measurements. In the second part of this paper, the stress effects on interfacial reliability of TSV structures are discussed. It is found that plastic deformation is highly localized in the Cu vias but can be sufficient to cause via extrusion without interfacial delamination. Alternatively, via extrusion by interfacial delamination is simulated by a shear-lag cohesive zone model that predicts the critical temperatures and critical TSV dimensions.

PRECISION WAFER CURVATURE TECHNIQUE

The precision wafer curvature technique is an extension of the wafer curvature technique that has been used extensively for stress measurement in thin films and periodic line structures [10-12]. The measurement system is set up based on an optical lever with a capability to measure the curvature to a precision of 6.5×10^{-5} m^{-1}. As shown schematically in Fig. 1a, the two incident laser beams are reflected by the specimen and the movement of the reflected laser spots is tracked by two position-sensitive photodetectors. The measurement system was designed with a heating stage inside a vacuum chamber. Therefore, the curvature change of the TSV specimen during thermal cycling can be measured *in situ* under a controlled atmosphere. More details of the system have been presented elsewhere [13].

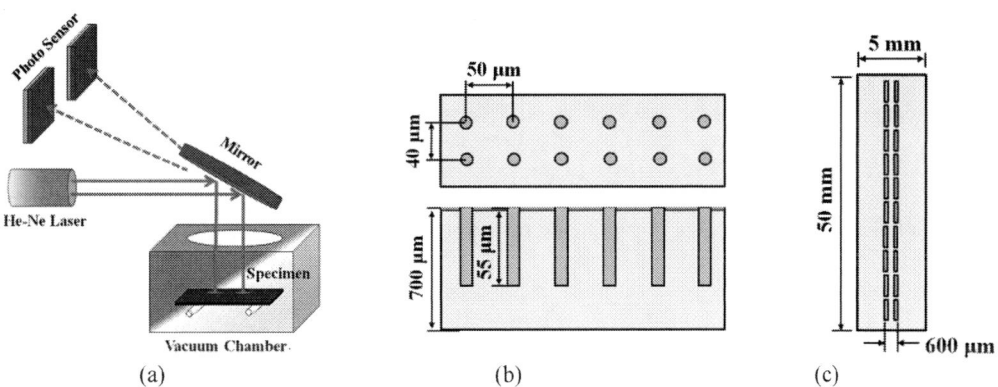

FIGURE 1. (a) Schematic of the precision wafer curvature measurement system. (b) Illustration of the TSV specimen, top and cross sectional views. (c) Top view of the TSV specimen for the curvature measurements, with TSV arrays in many blocks along the center line. Top view in (b) represents one block of arrays in (c).

The TSV structure used in the present study contains periodic arrays of blind Cu vias that are 10 μm in diameter with a nominal depth of 55 μm. The silicon wafer is 700μm thick and is of (001) type. The spacing between the TSVs is 40μm along the [110] direction and 50μm along the [1$\bar{1}$0] direction (Fig. 1b). For the curvature measurement, the wafer was cut into 5×50 mm beams where the TSVs were located along the centerline of the specimen (Fig. 1c). There was an oxide barrier layer of 0.4 μm thick at the via/Si interface and an oxide layer of 0.8 μm thick on the surface of the wafer. The surface oxide layer was mechanically removed for all measurements in this work.

The curvature measurements were conducted for several fully-filled TSV specimens subjected to different thermal cycling. To determine the residual stress in the Cu vias, a reference specimen was used by etching off the Cu vias. The curvature of the reference specimen was measured over the same thermal cycle as the specimen with fully filled Cu vias, and the curvature difference between the two specimens is attributed to the average thermal stress in the Cu vias. As shown in Fig. 2a, the curvature decreases nonlinearly with increasing temperature during the first cycle, suggesting an average compressive stress in the Cu vias and inelastic deformation. During cooling, however, the curvature changes linearly with the temperature, suggesting predominantly linear elastic deformation. In particular, the curvature difference between the two specimens becomes zero at around 100°C, suggesting a zero average stress in the Cu vias at this temperature. Below 100°C, the curvature becomes positive, and the average stress in the Cu vias becomes tensile. The temperature of zero curvature (~100°C) is taken as the reference temperature for subsequent thermal stress analysis.

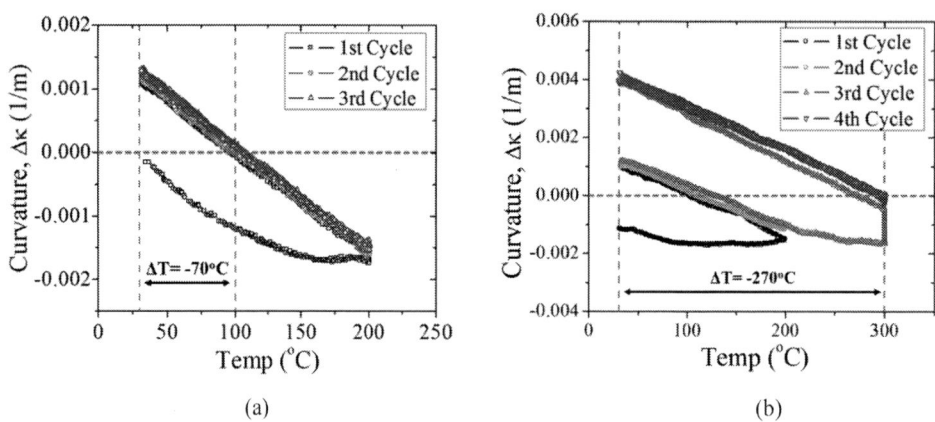

(a)   (b)

FIGURE 2. Curvature measurements for (a) a TSV specimen subjected to thermal cycling to 200°C; (b) a TSV specimen subjected to four thermal cycles with an annealing step at 300°C for 1 hour.

In the first measurement (Fig. 2a), the specimen went through three thermal cycles to 200°C at a heating rate of 2°C/min. After the first thermal cycle, the curvature-temperature relation became nearly linear and was reversible up to 200°C. The residual curvature at room temperature increased slightly after each cycle. In the second measurement (Fig. 2b), the specimen was heated to 200°C in the first two cycles, and then heated to 300°C during the third cycle and annealed for 1 hour prior to cooling, followed by an additional cycle to 300°C. For the first two cycles, the curvature behavior is similar to the first specimen. However, when the temperature was increased beyond 200°C during the third cycle, a nonlinear curvature-temperature behavior similar to the first cycle was observed from 200°C to 300°C. During annealing at 300°C, the curvature drops to almost zero. Evidently, the average stress in the Cu vias was relaxed considerably during annealing at 300°C. Subsequently, during cooling and the last thermal cycle, the curvature-temperature behavior again became nearly linear and reversible up to 300°C. Compared to Fig. 2a, the residual curvature in Fig. 2b is much larger after four thermal cycles, suggesting a higher tensile stress in the vias. This is attributed to the annealing process that reset the reference temperature to 300°C. Therefore, depending on the thermal history, different thermal load (ΔT) has to be used for the thermal stress analysis. In Fig. 2a, the reference temperature is 100°C, and the thermal load ΔT_A = -70°C at the room temperature (~30°C). For Fig. 4b, the reference temperature becomes 300°C after the annealing step, and the thermal load ΔT_B = -270°C. The reference temperatures determined here were used in the finite element analyses to compare with micro-Raman measurements, as discussed in Section 4.

The measured curvature-temperature behavior can be related to the average thermal stresses and the deformation mechanisms of the TSV specimen. The negative curvature indicates an average compressive stress in the Cu vias, while the positive curvature implies tensile stress. The nonlinear curvature-temperature behavior observed during the heating process of the first cycle suggests inelastic deformation mechanisms, which was found to be related to the evolution of the Cu grain structures, as discussed in Section 3 along with microstructure analysis. On the other hand, the nearly linear curvature-temperature behavior in the subsequent thermal cycles indicates predominantly linear elastic behavior of the Cu vias, which is in sharp contrast with the thermomechanical behavior of Cu thin films [12].

MICROSTRUCTURE ANALYSIS

To further understand the deformation mechanisms underlying the measured curvature-temperature behavior of the TSV specimens, microstructure evolution of the Cu vias subjected to different thermal histories was studied [6]. A number of TSV specimens were each subjected to a single thermal cycle to different temperatures, and the measured curvatures are shown in Fig. 3. Despite the different cycling temperatures ranging from 100°C to 400°C, similar behavior was observed for all specimens: a nonlinear curvature-temperature relation during heating followed by a nearly linear relation during cooling.

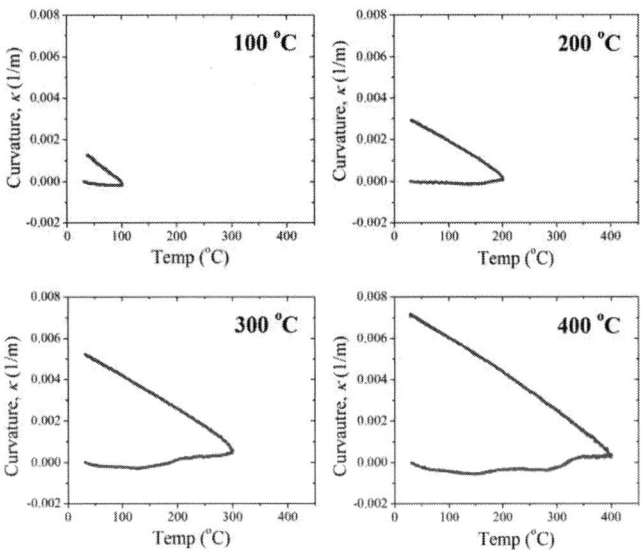

FIGURE 3. Curvature measurements for TSV specimens subject to thermal cycles with the highest temperature at 100°C, 200°C, 300°C, and 400°C.

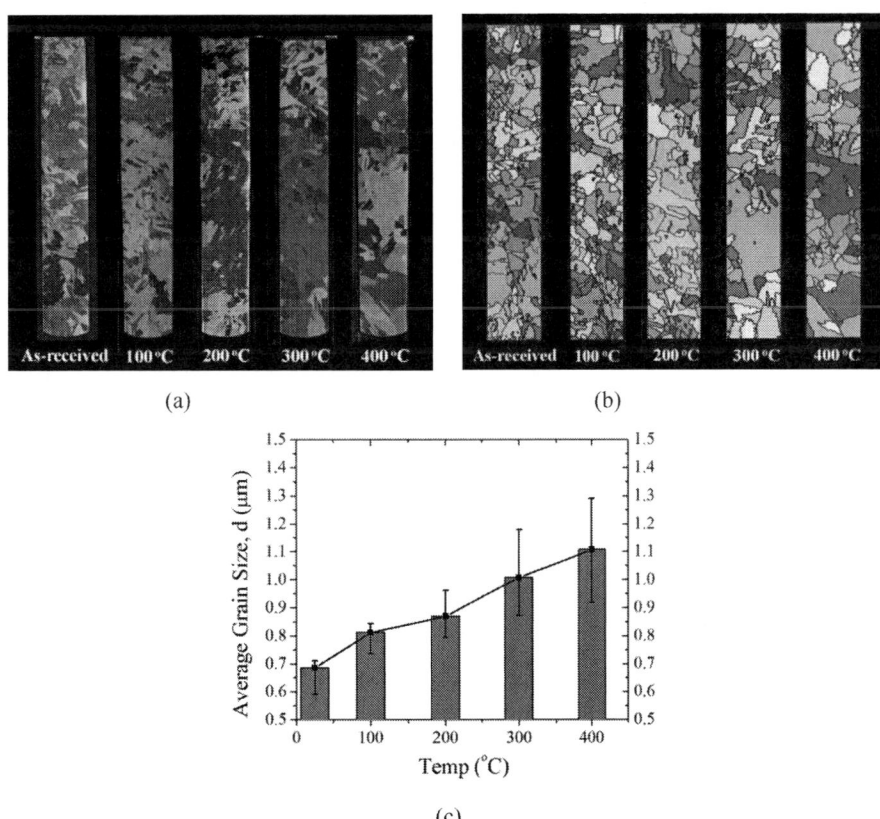

FIGURE 4. (a) Focused ion beam images of TSVs after different thermal loads. (b) Grain mapping by EBSD. (c) Average grain sizes.

Using focused ion beam (FIB), the cross-sections of the TSV specimens were examined after completing the thermal cycling measurements. The contrast of the ion channeling images of the Cu vias in Fig. 4a shows that the average Cu grain sizes are larger after thermal cycles and increase as the high end of the thermal cycling temperature increases, suggesting possible grain growth during the thermal cycles. This is confirmed by electron backscatter diffraction (EBSD) analysis of the grain structures. The EBSD grain mappings for the Cu vias are shown in Fig. 4b together with the average grain sizes measured and compared in Fig. 4c. Evidently, systematic grain growth has occurred in the Cu vias after each thermal cycle. The average grain size for the via in the as-received TSV specimen is 0.69 μm. After thermal cycling to 100, 200, 300, and 400°C, the average grain sizes have grown by 18.4%, 26.8%, 46.8%, and 61.4%, to 0.81, 0.87, 1.00, and 1.11 μm, respectively.

With the EBSD technique, the grain orientation of the Cu vias was quantitatively measured. In Fig. 5, the inverse pole figures of the grain orientations are plotted for the TSVs along the directions normal to the TSV length (ND) and parallel to the TSV length (RD). Overall, there appears to be no preferred Cu grain orientation in all the specimens before and after thermal cycling. The lack of preferred grain orientation indicates a statistically isotropic grain structure in the Cu vias, and thus the thermomechanical properties of the Cu can be treated as isotropic in the thermal stress analysis. In addition, the misorientation across grain boundaries obtained from the EBSD measurements is plotted in Fig. 6. There exist a large number of twin boundaries with a characteristic misorientation angle of 60° across the grain boundaries for all the vias examined. The presence of twin boundaries may lead to relatively high yield strength of the Cu vias.

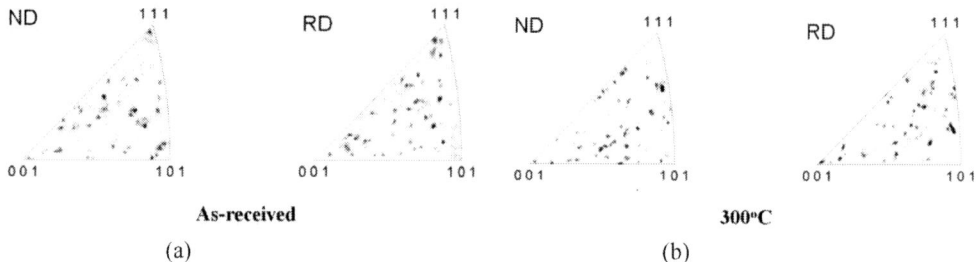

(a)   (b)

FIGURE 5. Inverse pole figures of (a) as-received TSV and (b) TSV after thermal cycling to 300°C. Two measurement directions were defined: ND (normal to the TSV axis) and RD (parallel to the TSV axis).

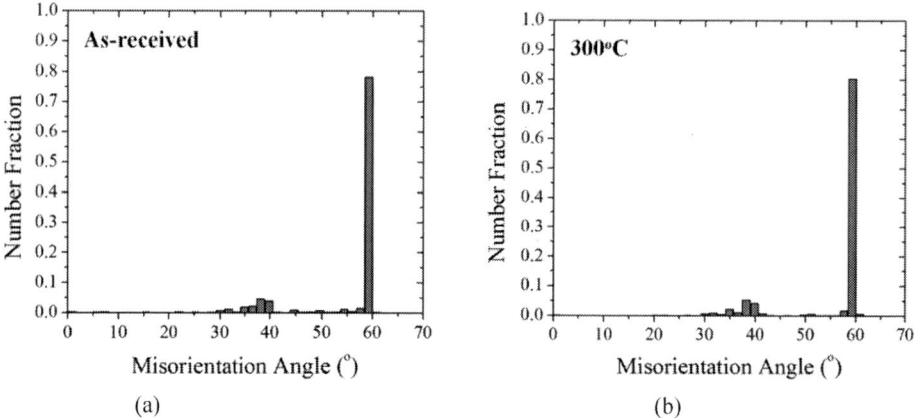

(a)   (b)

FIGURE 6. Grain misorientation angles obtained by EBSD for (a) as-received TSV and (b) TSV after thermal cycling to 300°C.

Based on the microstructure analysis, the curvature-temperature behavior of the fully-filled TSV specimen can be understood as the following. The nonlinear curvature-temperature relation during heating of the first cycle is mainly attributed to the nonlinear stress relaxation caused by grain growth. Similar curvature behavior due to grain growth has been observed for Cu thin films [14,15]. As grain growth proceeds to eliminate grain boundaries and reduce the excess volume, it is favored when the average stress in the Cu vias is compressive during heating [16]. The nearly linear curvature-temperature behavior during cooling and subsequent cycles suggests stabilized grain structures in the Cu vias. The grain structures would remain stabilized as long as the temperature does not exceed the highest temperature that the TSV specimen has experienced in any of the previous cycles. When the temperature increased beyond the highest temperature in the previous cycles, the grain structures would evolve further with additional grain growth and stress relaxation. Furthermore, the annealing process in Fig. 2b shows continual stress relaxation at the high temperature. In general, grain growth is a kinetic process that depends on both temperature and stress.

MICRO-RAMAN SPECTROSCOPY

For the TSV structure, the thermal stresses in Cu can in turn induce stresses in the Si matrix surrounding the TSVs where the stress distribution near the wafer surface is particularly important since most of the active devices are located near the surface. To measure the near surface stresses in Si, the micro-Raman spectroscopy technique was used [9]. Raman spectroscopy relies on the inelastic scattering (or Raman scattering) of Si, where the frequency shift of the Raman modes provides a measure of the stress in Si. The theory of Raman measurement and its application for TSV structures have been developed previously [17]. Under the [001] backscattering configuration, only the longitudinal Raman mode can be detected. Assuming a biaxial stress state near the wafer surface, the following relation can be deduced from the secular equation for (001) Si [18],

$$\sigma_r + \sigma_\theta \ (MPa) = -470\Delta\omega_3 (cm^{-1}),\tag{1}$$

where $\sigma_r + \sigma_\theta$ is the sum of the in-plane normal stresses, and $\Delta\omega_3 = \omega_3 - \omega_0$ is the Raman frequency shift of the longitudinal Raman mode. With Eq. (1), the stress sum near the wafer surface can be determined from the measurement of $\Delta\omega_3$.

In this study, Raman measurements were carried out with a commercial micro-Raman Spectrometer equipped with a 442 nm Ar laser. Two TSV specimens were subjected to similar thermal treatment as those in the wafer curvature experiments (Fig. 2). Specimen A was heated to 200°C and then immediately cooled down to room temperature (RT), and specimen B was heated to 300°C and annealed for 1 hour prior to cooling down. For both specimens, the Raman measurements were conducted at RT by scanning across two neighboring vias along the [110] direction. To deduce the frequency shift $\Delta\omega_3$, a reference Raman frequency ω_0 is required, which was determined by extending the measurement to areas far away from the TSVs where the stress is assumed to be zero. With the calibrated reference frequency ω_0, the sum of the two principal stresses in Si is deduced from the measured Raman frequency using Eq. (1).

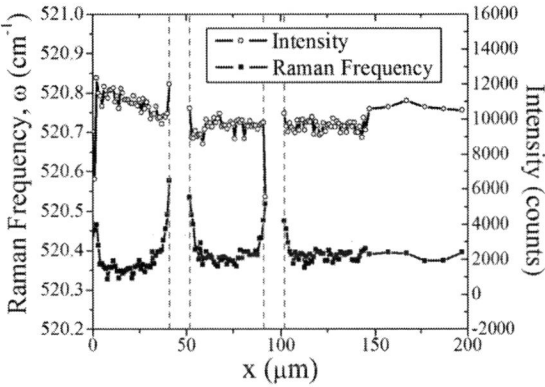

FIGURE 7. Raman intensity (open symbols) and frequency (filled symbols) of a TSV specimen. Dashed lines indicate the Cu/Si interfaces.

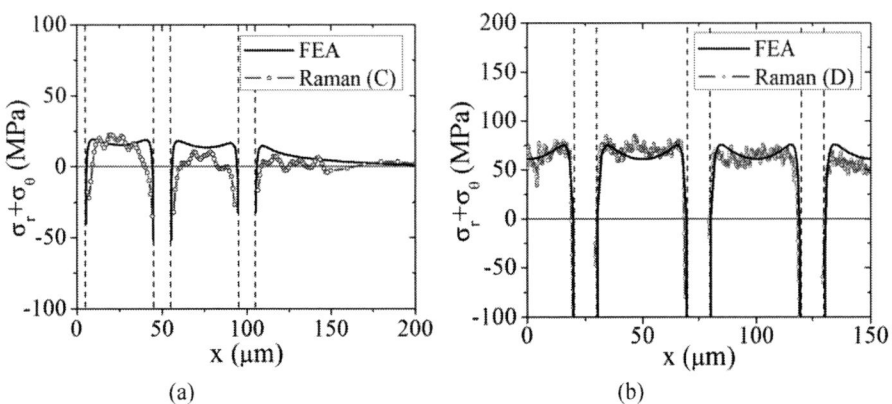

(a)   (b)

FIGURE 8. Comparison of the near-surface stress distribution between Raman measurements and FEA: (a) Specimen A; (b) Specimen B.

The measured Raman intensity and frequency shift obtained from specimen A are shown in Fig. 7, representing typical results obtained from Raman measurements. A sudden drop of the Raman intensity was observed near the Cu/Si interface. The distributions of the stress sums deduced from the Raman measurements are plotted in Fig. 8 for both specimens. Clearly, a sub-micron resolution was achieved in the measurement, but the results provided only the sum of the two individual stress components in Si. Further understanding of the stress characteristics in the TSV structure requires detailed stress analysis to delineate the stress components and correlate the micro-Raman measurements with the thermal cycling experiments [6, 9].

A three-dimensional finite element model was constructed using the commercial package, ABAQUS (v6.10). A quarter of the via with symmetric boundary conditions in the [110] and [1$\overline{1}$0] directions was modeled to simulate the periodic TSV array used in the Raman measurement. The anisotropy of Si was taken into consideration by using the anisotropic elastic constants for Si, and Cu is treated as isotropic based on the microstructure analysis by EBSD. The following material properties were used for Cu and SiO_2: Young's modulus, $E_{Cu} = 110$ GPa and $E_{oxide} = 70$ GPa; Poisson's ratio, $v_{Cu} = 0.35$ and $v_{oxide} = 0.16$. The CTEs are $\alpha_{Cu} = 17$ ppm/$^\circ$C, $\alpha_{Si} = 2.3$ ppm/$^\circ$C and $\alpha_{oxide} = 0.55$ ppm/$^\circ$C. Since the Raman signal penetrates up to 0.2 μm from the wafer surface, the stress components are extracted from 0.2 μm below the wafer surface. The sums of the in-plane stresses obtained by FEA for specimens A and B are calculated and compared with the Raman measurements. The thermal loads for the two specimens were chosen to be the same as those in the curvature measurements (Fig. 2) to facilitate the correlation of the results from the two techniques. Based on the curvature measurements, the reference temperature for specimen A is taken to be 100°C, and that for specimen B is 300°C, corresponding to thermal loads of $\Delta T_A = -70^\circ$C and $\Delta T_B = -270^\circ$C, relative to the room temperature of 30°C. As shown in Fig. 8, the FEA results are in reasonable agreements with the Raman measurements. Moving away from the Cu/Si interface, the sum of the stresses first increases sharply, and then gradually decreases. Between the two adjacent vias, the stress depends on the pitch distance as a result of the stress interaction. The measurement for specimen B (Fig. 8b) shows a higher stress level in Si than for specimen A, as a result of the higher negative thermal load ($|\Delta T_B| > |\Delta T_A|$). Therefore, the stresses in Si around the TSVs depend on the thermal processes of the specimen.

INTERFACIAL RELIABILITY

Effect of Cu Plasticity on Via Extrusion

After thermal cycling, via extrusion was observed in the TSV specimen (Fig. 9a). A previous study has suggested that via extrusion could be caused by interfacial delamination [19]. However, in the present study, no evidence of interfacial delamination was observed. Instead, via extrusion appears to have occurred as a result of localized plastic deformation near the via/Si interface during thermal cycling. An elastic-plastic FEA model was constructed to investigate the effect of Cu plasticity on via extrusion. In general, the plastic deformation in the Cu TSVs depends on the thermal load and the yield strength of Cu. For the present study, the yield strength of Cu was assumed to be 300 MPa and a thermal load of $\Delta T = 270^\circ$C was applied. In Fig. 9b, the deformed shape by the FEA model clearly showed extrusion of the via, similar to what was observed in our experiments. It is noted that plastic deformation in the Cu via is highly localized, as shown in Fig. 9c, where the equivalent plastic strain in Cu is non-zero only in a small region near the top of the via. The plastic yielding of Cu near the interface effectively relaxes the constraint of the surrounding materials and allows the via extrusion without interfacial delamination. Moreover, the local plasticity in Cu could also enhance the total fracture energy for interfacial delamination [20] and thus help prevent delamination.

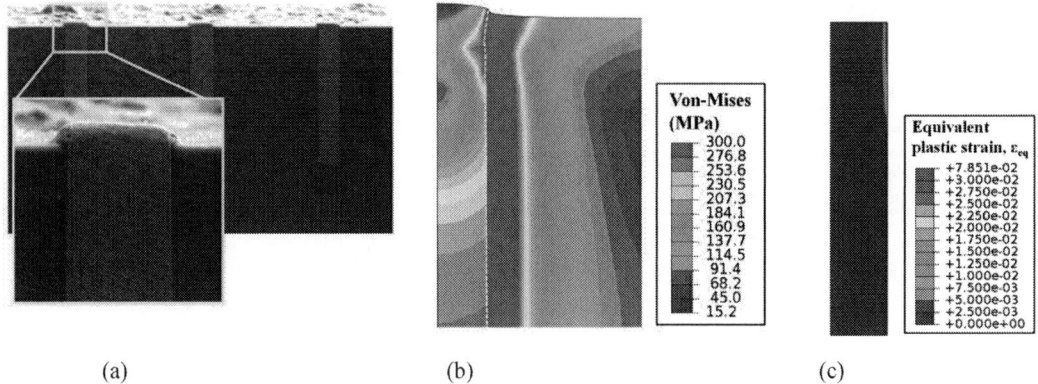

(a) (b) (c)

FIGURE 9. (a) SEM image of TSV extrusion observed after thermal cycling. (b) Stress distribution and deformation of TSV by an elastic-plastic FEA model. (c) Equivalent plastic strain in the TSV by FEA (yield strength = 300 MPa, $\Delta T = 270^{\circ}$C).

Via Extrusion by Interfacial Delamination

An alternative mechanism for via extrusion is due to interfacial delamination as observed in some studies [21]. To predict the critical condition for initiation of interfacial delamination, a cohesive zone model was adopted [22]. Using a bilinear traction-separation law for the interface, the interface first undergoes elastic deformation until the combination of the opening stress (mode I) and shear stress (mode II) reaches a critical level, which depends on the cohesive strength of the interface. Subsequently, the interface is partly damaged and weakened upon further loading. A delamination crack is nucleated when a critical separation (both opening and shearing) is reached, which depends on the fracture toughness of the interface. Therefore, two critical conditions can be determined, one for damage initiation and the other for crack nucleation. For the TSV structure subject to a positive thermal load ($\Delta T > 0$), the interfacial delamination is predominantly mode II (shearing), which can be analyzed by a shear-lag model [22]. For given material properties and via dimensions, two critical temperatures are predicted. For damage initiation, the critical temperature is

$$\Delta T_{c1} = \frac{2}{(\alpha_{TSV} - \alpha_{Si})}\sqrt{\frac{\tau_i \delta_i}{E_{TSV}D}}\coth\left(\frac{H}{\lambda}\right) \tag{2}$$

where $\lambda = \sqrt{\dfrac{E_{TSV}D\delta_i}{\tau_i}}$ is a characteristic length scale, τ_i is the shear strength of the interface, δ_i is the critical

separation, E_{TSV} is Young's modulus of the via material, D is via diameter, H is via height, α_{TSV} and α_{Si} are the coefficients of thermal expansion. For crack nucleation, the critical temperature is

$$\Delta T_{c2} = \frac{2}{(\alpha_{TSV} - \alpha_{Si})}\sqrt{\frac{2\Gamma_i}{E_{TSV}D}} \tag{3}$$

where Γ_i is the fracture toughness of the interface.

As an example, we plot in Fig. 10 the two critical temperatures versus the via diameter, taking the following material properties: E_{TSV} = 110 GPa, α_{TSV}-α_{Si}=14.7 ppm/$^{\circ}$C, $\Gamma_i = 10.0$ J/m^2, τ_i=300 MPa, and δ_i=20 nm. When the thermal load (ΔT) is specified for TSV processes, the critical via diameters can be determined. In general, vias with larger diameters are more prone to interfacial delamination and hence via extrusion. The model prediction agrees qualitatively with the experiments where via extrusion was observed for via diameters greater than a critical value [21], while the dependence on the via height is much weaker.

155

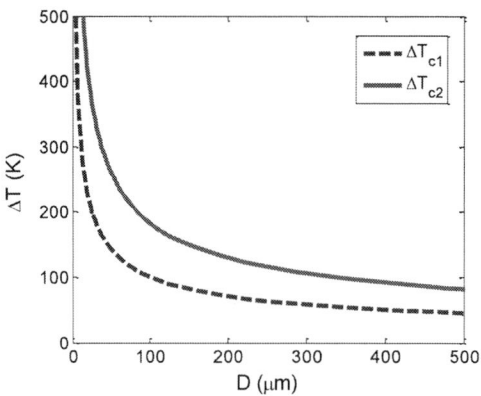

FIGURE 10. Critical temperatures predicted by the shear-lag cohesive zone model.

SUMMARY

This paper presents thermomechanical characterization of TSV structures by combining the precision wafer curvature technique in thermal cycling experiments with micro-Raman spectroscopy. The evolution of the Cu microstructure in the vias was analyzed by focused ion beam (FIB) and electron backscatter diffraction (EBSD) techniques, which provided insights into the underlying mechanisms for elastic and inelastic deformation in the TSV structures. Via extrusion was observed after thermal cycles, which may be due to local plastic deformation or interfacial delamination. Finite element analysis showed that plastic deformation is highly localized near the via/Si interface, which could be sufficient to cause via extrusion without interfacial delamination. Alternatively, a shear-lag cohesive zone model was developed to predict the critical temperatures for interfacial damage initiation and crack nucleation towards delamination.

ACKNOWLEDGMENTS

This work was supported by Semiconductor Research Corporation.

REFERENCES

1. N. Ranganathan, K. Prasad, N. Balasubramanian, and K. L. Pey, *J. Micromech. Microeng.* 18, 075018 (2008).
2. C.S. Selvanayagam, J.H. Lau, X.Zhang, S. Seah, K. Vaidyanathan, and T.C. Chai, *IEEE Trans. Advanced Packaging* 32, 720-728 (2009).
3. A. P. Karmarker, X. Xu, and V.Moroz, *Proc. IEEE 47th Int. Reliab. Phys. Symp.* (Montreal, Canada), pp. 682–687 (2009).
4. K.H. Lu, X.F. Zhang, S.K. Ryu, J. Im, R. Huang, and P.S. Ho, *Proc. IEEE 59th ECTC* (San Diego, CA), pp. 630–634 (2009).
5. S.K. Ryu, T. Jiang, K.H. Lu, J. Im, H.-Y. Son, K.-Y. Byun, R. Huang, and P.S. Ho, *Appl. Phys. Lett.* 100, 041901 (2012).
6. T. Jiang, S.K. Ryu, Q. Zhao, J. Im, R. Huang and P.S. Ho, *Microelectronics Reliability*, 53, pp. 53-62 (2013).
7. S.K. Ryu, K.H. Lu, X. Zhang, J.H. Im, P.S. Ho and R. Huang, *IEEE Trans. Device and Materials Reliability* 11, 35-43 (2011).
8. S.K. Ryu, K.H. Lu, T. Jiang, J. Im, R. Huang, and P.S. Ho, *IEEE Trans. Device and Materials Reliability* 12, 255-262 (2012).
9. S.K. Ryu, Q. Zhao, M. Hecker, H.-Y. Son, K.-Y. Byun, J. Im, P.S. Ho, and R. Huang, *J. Appl. Phys.* 111, 063513 (2012).
10. R.P. Vinci, E.M. Zielinski and J.C. Bravman, *Thin Solid Films* 262, 142-153 (1995).
11. I.-S. Yeo, P.S. Ho, and S.G.H. Anderson, *J. Appl. Phys.* 78, 945 (1995).
12. D. Gan, P.S. Ho, R. Huang, J. Leu, J. Maiz and T. Scherban, *J. Appl. Phys.* 97, 103531 (2005).
13. I.-S. Yeo, *Thermal stresses and stress relaxation in Al-based metallization for ULSI interconnects*, Ph.D. thesis, University of Texas at Austin, 1996.
14. D. C. Miller, C. F. Herrmann, H. J. Maier, S. M. George, C. R. Stoldt and K. Gall, *Thin Solid Films* 515, 3208-3223 (2007).
15. J.M.E. Harper, C. Cabral, P.C. Andricacos, L. Gignac, I.C. Noyan, K.P. Rodbell, and C.K. Hu, *J. Appl. Phys.* 86, 2516-2525 (1999).

16. P. Chaudhari, *J. Vac. Sci. Tech.*, 9, 520-522 (1972).
17. I. De Wolf, H. E. Maes, and S. K. Jones, *J. Appl. Phys.* 79, 7148-7156 (1996).
18. M. Hecker, L. Zhu, C. Georgi, I. Zienert, J. Rinderknecht, H. Geisler, and E. Zschech, *AIP Conf. Proc.* 931, pp. 435-444 (2007).
19. S.K. Ryu, K. Lu, J. Im, R. Huang and P.S. Ho, *AIP Conf. Proc.* 1378, pp. 153-167 (2011).
20. M. Lane, R.H. Dauskardt, A. Vainchtein, and H.J. Gao, *J. Mater. Res.* 15, 2758-2769 (2000).
21. S. Cho, RTI 3D Workshop (Burlingame, CA), December 2010.
22. S.K. Ryu, *Thermomechanical Stress Analysis and Interfacial Reliability for Through-Silicon Vias in Three-Dimensional Interconnect Structures.* PhD dissertation, University of Texas at Austin, 2011.

Assessment Of Fracture And Elastoplastic Properties Of Thin And Very Thin Films

M. Trueba[1], D. Gonzalez[1], I. Ocaña[1], M.R. Elizalde[1*], J.M. Martinez-Esnaola[1], M.T. Hernandez[2], M. Haverty[2], G. Xu[3], D. Pantuso[2]

[1]CEIT and TECNUN (University of Navarra), Manuel Lardizabal 15, 20018 San Sebastián, Spain
[2]Design Technology Solutions, Intel Corporation, Hillsboro 97124 (OR), USA
[3]Logic Technology Q&R, Intel Corporation, Hillsboro 97124 (OR), USA

Abstract. Microelectronic industry is driven by the continuous miniaturization process conducing to the introduction of new materials. These materials are subjected to stresses mainly due to thermal mismatch, microstructural changes or process integration which can be in the origin of mechanical reliability issues. To study these phenomena and even electromigration a good mechanical characterization of the materials is needed. This work aims at developing tests to assess fracture or elastoplastic behavior of thin and very thin metallic and polymeric films. The tests developed are based on indentation combined with sample preparation using mostly FIB. Among other techniques, different test geometries for microbeams have been evaluated and quantitative data have been obtained combining experimental results with analytical or numerical models, depending on the material under study. For the particular case of Cu cantilevers, a strong dependence of their plastic behavior on the orientation of the grains close to the fixed end has been detected. Grain orientation has been measured by EBSD and the plastic behavior has been modeled by FEM using an in-house crystal plasticity subroutine.

Keywords: thin films, microbeams, strength, toughness
PACS: 68.60.Bs; 46.35.+z

INTRODUCTION

As the microelectronic industry advances, the materials used in integrated circuits are improved from their electronic and optic point of view. But the mechanical properties cannot be disregarded as residual stresses that appear as a consequence of the deposition processes, stresses during packaging and the in-service thermal cycling can make cracks grow and compromise the reliability of the chip.

Thin films are used in many technologies in the microelectronic industry. The traditional mechanical testing techniques are not enough to measure the mechanical properties of the thin films. Many different techniques have been developed to characterize their properties and the properties of their interfaces.

As for characterizing interfaces, two techniques are mainly used in the industry. Four point bending (developed by Dauskardt et al. [1]) is the reference testing method in the industry. In this test, a macroscopic notched sample (10 mm long) is subjected to bending to initiate a crack which kinks and propagates along the interface of interest. The main limitations of this technique are that (i) special specimens involving adhesives are required, thus increasing the throughput time and introducing variability in the results, and (ii) the cracks generated are of the order of millimeters and hence local properties cannot be determined (drawback for patterned structures). As an alternative method, Sánchez et al. [2] developed the cross-sectional nanoindentation (CSN), that was further developed and adapted for characterizing pattern films by Ocaña et al. [3]. In CSN a crack is initiated in the silicon underlying the structures of interest by nanoindentation, and the crack propagation along different interfaces is measured and used to characterize their adhesion energy.

Concerning the fracture characterization of thin films, several efforts have been made. The technique mainly used for this purpose in industry is "channel cracking". Developed by Huang et al. [4] in channel cracking a crack initiated from a scratch is propagated by bending the sample. The film fracture energy is calculated from the stress needed for crack propagation. Even though it is the reference test in industry, this technique gives quite random results and has been proven as extremely sensitive to the operator. Indentation techniques have also been used to measure fracture toughness. A sharp tip (typically a Vickers, a Berkovich or a cube corner diamond) is pushed into bulk brittle materials. If the applied load reaches the critical value, radial cracking can occur. Using the maximum load and the crack length, fracture toughness can be calculated [5,6]. These techniques have been developed in order to obtain the fracture toughness of ceramic materials [7]. Finite element modeling is necessary to describe the complex stress and strain fields that appear under the indenter tip [8]. The main drawback of these techniques is that the crack patterns obtained depend on the system tested (thin film thickness, substrate properties, residual stresses, interfacial properties) which reduces the reproducibility and makes quantification very challenging. Testing of microsamples has been widely used to characterize mechanical properties of small volumes. For instance Matoy et al. [9] used bending of cantilever beams (machined by etching processes) to calculate the fracture toughness of silicon based dielectric materials and found that it increased with decreasing cantilever thickness.

In this work, the ability to obtain fracture properties of metallic and polymeric thin films testing microbeams has been explored. An alternative method to machine samples using FIB is presented. The beams have been tested using a nanoindenter. Cantilever beams have been tested up to fracture. In the case of brittle materials, the maximum load together with the geometrical parameters and the Young's modulus of the materials are used to calculate the fracture stress of the thin films. For ductile materials, the energy introduced during the test and the cracked area are also needed to calculate the fracture toughness. Finally, for the particular case of Cu cantilevers a methodology to determine the plastic behavior of individual Cu grains is presented.

MATERIALS AND EXPERIMENTAL TECHNIQUES

The samples studied in this work are polymeric and metallic blanket thin films deposited on a {111} Si wafer. Table 1 shows the materials and stacks that have been tested. A Young's modulus of 155 GPa is assumed for the Si.

The beams have been machined at the FIB. Sample preparation starts cleaving small pieces of about 10 mm × 10 mm by pre-cracking one edge of the sample with a scriber and propagating the crack by bending the sample with special pliers.

TABLE 1. Materials Tested.

Material	Stack	Young's Modulus (GPa)
Polymer	750 nm polymer on Si	10
Cu_A	Cu (0.5 μm) – SiOx (100 nm) – TNT (10 nm)-Si	-
Cu_B	Cu (1 μm) – SiOx (100 nm) – TNT (10 nm)-Si	-

These pieces are introduced in the FIB chamber (QUANTA3DFEG, FEI) and different microbeam geometries are machined using the following general procedure. The sample is stuck on a holder and introduced in the chamber with the top surface perpendicular to the electron beam. The sample is tilted 52° so the surface is perpendicular to the ion beam and the top geometry is defined (see figure 1). Positioning marks, which will help the positioning of the tip when testing each beam at the nanoindenter, are also machined in this step.

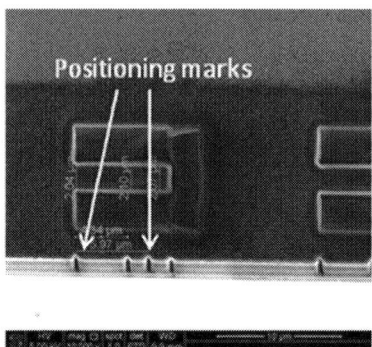

FIGURE 1. SEM images showing details of the first step of the machining process for cantilever beams. Some typical dimensions are identified for the numerical analysis.

Then, the sample is taken out of the chamber and is rotated 90°, so the top surface is now parallel to the electron beam. To define the thickness of the beam and allow its deflection a trench is machined in cross section. This trench is milled at a distance h from the top surface, determining the thickness of the beam. Figure 2 shows the typical geometry obtained using the described method.

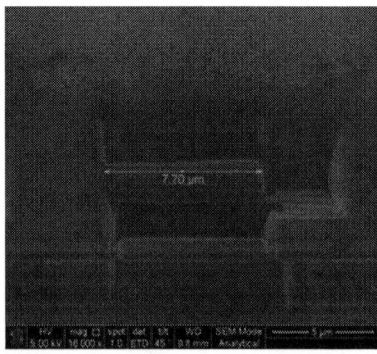

FIGURE 2. SEM image taken at 45° showing the typical geometry obtained using the described method for cantilever beams.

The beams are deflected at the TriboIndenter[TM] (Hysitron, USA) using a conical tip (tip radius 1.86 μm) until fracture occurs. This point can be detected as a sudden jump in displacement in the load – displacement record.

The testing process is as follows. First, an SPM (Scanning Probe Microscopy) image of the beam is taken with the imaging mode of the TriboIndenter[TM] (figure 3). In this mode the center of the image is the point where the load is applied, characterized by the distance to the fixed end ($L/2$ for the case of beams fixed at both ends and L for cantilever beams). All the tests are carried out under displacement control. After testing, a fractographic analysis is performed at the FIB and the crack initiation sites and/or propagation paths are analyzed.

FIGURE 3. Images taken with the imaging mode showing the typical aspect of a cantilever beam.

RESULTS AND DISCUSSION

Polymer

As it will be explained later, to increase the stiffness of the system and make it possible to achieve the fracture point, composed beams with approximately 2.5 μm of Si supporting the polymeric film have been machined. This configuration leads to successful tests for this material because a crack starts at the polymer and propagates through all the thickness of the beam. Figure 4 shows an example of a polymer-Si beam after testing and a typical load-displacement record.

When working with very thin films, if a simple beam is machined (with the thickness limited by the total one of the film) the resulting geometry is often too compliant and the machine is not able to detect the surface, breaking the sample without any load–displacement record. This problem was overcome by machining composed beams (polymer + Si). With a simple analytical analysis of a cantilever beam, two points in which fracture can start can be identified (figure 5): A (maximum stress in the ILD) and B (maximum stress in the Si).

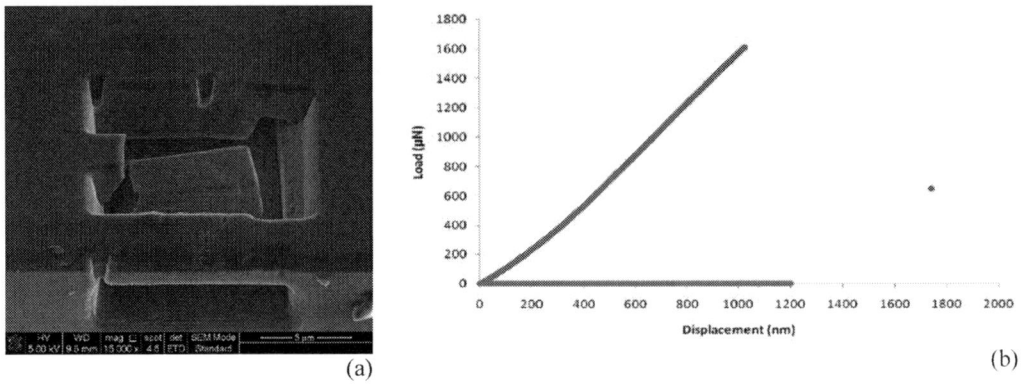

(a)   (b)

FIGURE 4. Test performed on a polymer-Si cantilever: (a) SEM image showing a fractured cantilever after a test and (b) *t*ypical load–displacement record.

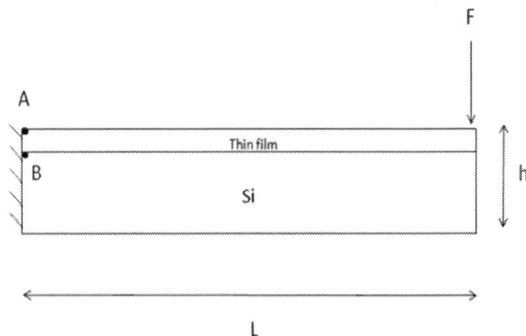

FIGURE 5. Sketch of a test in a cantilever beam with the 2 points with maximum stresses identified (A for the ILD and B for the Si).

The stresses in A and B are calculated using the bending theory of beams as

$$\sigma_A = \frac{F \cdot L \cdot z_A}{I_{eq}} \cdot \frac{E_1}{E_2} \qquad (1)$$

$$\sigma_B = \frac{F \cdot L \cdot z_B}{I_{eq}} \qquad (2)$$

where F is the applied load, L the distance between the fixed end and the point at which the load is applied, z is the distance to the neutral axis, I_{eq} is the equivalent moment of inertia with respect to the transversal axis and E_1 and E_2 are the Young's modulus of the thin film and the Si, respectively.

The crack should start in the ILD if the mechanical properties of the ILD are to be evaluated. The condition for this is that the relation between the fracture strength of the thin film and the Si must fulfill:

$$\sigma_{fracture,Si} > \sigma_B \tag{3}$$

when

$$\sigma_A = \sigma_{fracture,ILD} = \frac{F \cdot L \cdot z_A}{I_{eq}} \cdot \frac{E_1}{E_2} \tag{4}$$

An expression of F as a function of $\sigma_{fracture,ILD}$ can be derived and substituted in (2), which leads to the following ratio between the strength of the Si and the ILD to have fracture starting at the ILD:

$$\frac{\sigma_{fracture,Si}}{\sigma_{fracture,ILD}} > \frac{z_B}{z_A} \cdot \frac{E_2}{E_1} \tag{5}$$

In the tests performed, $\dfrac{z_B}{z_A} \approx 0.6$ and $\dfrac{E_2}{E_1} \approx 15.5$, so

$$\sigma_{fracture,ILD} < \frac{\sigma_{fracture,Si}}{9.3} \tag{6}$$

Using these analytical calculations together with the maximum load obtained in the experiments, the dimensions of the beams and assuming a Young's modulus of 10 GPa for the polymer, the following fracture strength of the polymer is obtained:

$\sigma_{fracture,ILD}= 560 \pm 100$ MPa

Assuming a fracture strength of 7000 MPa for the Si, equation (6) is fulfilled. So the crack is starting at the polymer, as desired.

Copper

Following the methodology described above, a first attempt with composed cantilever beams was made in Cu samples. Due to the ductility of the Cu, the crack starts at the Si, making it very difficult to extract any meaningful result out of this kind of test. Figure 6 shows a Cu-Si beam with the crack starting at the Si, illustrating this problem.

Inducing a stress concentration in the beams (for instance through notching) was envisioned as a possible approach for ductile materials. The pre-notch induces high stresses in the beam and forces the crack to start at this place. In this way repetitivity in the crack initiation site is obtained.

FIGURE 6. SEM image showing a cantilever beam of Cu-Si with a crack starting at the Si.

The thickness of the beam is determined by the thin film and the pre-notch further reduces the stiffness of the beam. In order to increase the stiffness of the beams without any Si support at all, the beams have been tested applying the load at the cross-section. The typical test configuration is shown in figure 7.

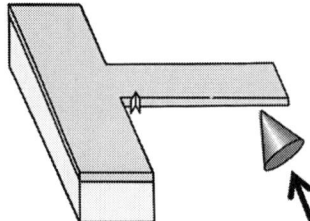

FIGURE 7. Sketch of a pre-notched cantilever beam with Cu on top and tested from the cross-section.

Figure 8 shows a pre-notched cantilever of 1 μm Cu, before and after the test. The crack has propagated through the notch, as desired. A typical load–displacement record for this kind of tests is shown in figure 9.

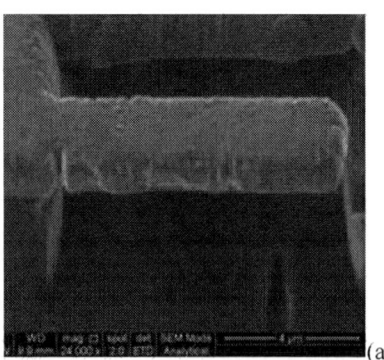 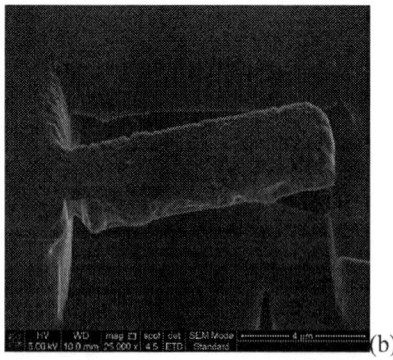

FIGURE 8. SEM image showing a pre-notched cantilever beam of 1 μm Cu a) before and b) after testing.

As a first attempt to characterize the fracture behavior of ductile thin films, plotting a "resistance curve" is proposed. Figure 10 shows the evolution of the total energy absorbed during the test with the propagated crack area for the cases of 0.5 and 1 μm Cu thick pre-notched cantilevers.

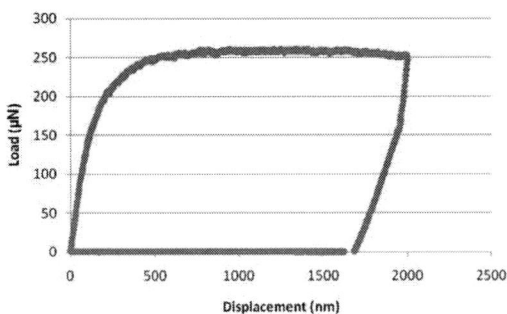

FIGURE 9. Load–displacement record of a 1 μm Cu pre-notched cantilever.

In these plots the propagation of the crack was evaluated making several cuts with the ion beam once the test was done. This way, the total cracked area can be measured. On the other hand, the energy introduced during the test was evaluated as the area below the load–displacement curve.

A further effort to rationalize the data obtained for ductile materials can be made taking into account the contribution of the plastic deformation to the total energy in the test (equation (7)),

$$W = W_f \cdot \Delta a \cdot h + W_p \cdot Fa^2 \cdot h + W_b \cdot h \qquad (7)$$

where W is the total energy in the test , $W_f \cdot \Delta a \cdot h$ is the energy spent in the fracture propagation and $W_p \cdot Fa^2 \cdot h + W_b h$ accounts for the plastic energy spent during the crack propagation including the blunting of the crack tip.

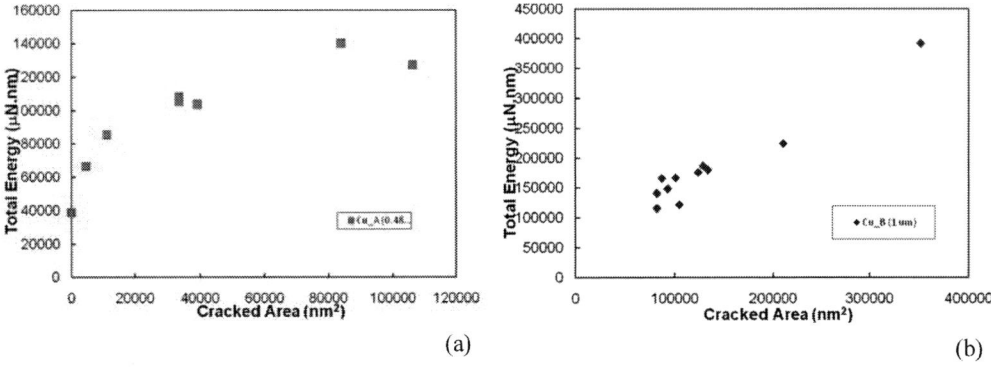

FIGURE 10. Total energy vs. cracked area for pre-notched cantilevers with (a) 0.5 μm thick Cu and (b) 1 μm thick Cu.

Taking into account equation (7) from the experimental data the fracture energy of the Cu films studied can be obtained:

$W_{f\,Cu\,-\,1\mu m} = 1\ kJ/m^2$

$W_{f\,Cu\,-\,0.5\mu m} = 0.7\ kJ/m^2$

In order to study the elastoplastic behavior of 1 μm Cu thin film, beams were machined as described for brittle materials (without notches) and tested from the top. The beams had a thickness smaller than 1 μm and hence were made only of Cu. The load-displacement curves showed dispersion which could not be explained in terms of differences in dimensions of the beam. It was observed that the Cu thin film had a columnar grain structure and that the beams had one or two grains in width. Grain orientation was measured by EBSD in a Jeol JSM-7000F (JEOL Ltd., Tokyo, Japan) equipped with an HKL system (fig 11a). This data was used to generate an FEM model of the beam including the columnar grain structure as shown in figure 11b. The beam test has been modeled using Abaqus. The material behavior is introduced using a user subroutine (UMAT) developed in-house which incorporates the strain gradient dependent crystal plasticity [10].

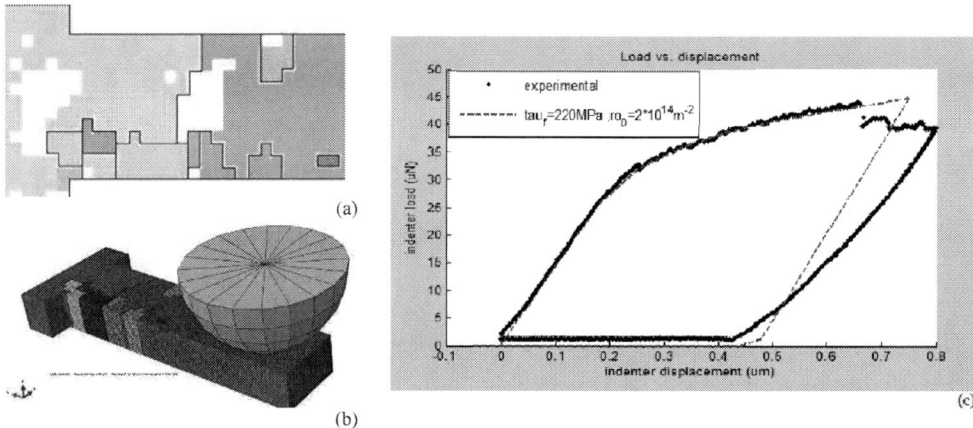

FIGURE 11. Characterization through FEM simulation of a Cu thin film from a bending test of a cantilever machined using FIB. (a) Orientation of Cu grains close to the fixed end measured by OIM (b) FEM simulation model including the grain structure. An in-house crystal plasticity subroutine is used. (c) Experimental and simulated load-displacement curve.

Cu properties are obtained fitting experimental results. The fitting parameters are the initial dislocation density, which in this case is fixed and has a value of $2\times10^{14}\ m^{-2}$, and the friction stress τ_f, which is related to the critical resolved shear stress. Figure 11c shows an experimental load-displacement curve and the corresponding simulated curve. The best fit is obtained for a τ_f value of 220 MPa, which is within the range from 90 to 270 MPa reported for fcc metallic micro- and nano-pillars tested in compression and tension [11].

166

CONCLUSIONS

The ability of microbeam testing to obtain fracture properties in very thin films has been demonstrated and different setups have been proposed. The technique has been developed and tested for polymeric and metallic thin films. Even though the technique is best suited to calculate fracture stress in brittle films, meaningful results can be obtained for ductile materials (such as metals) using notched beams.

A methodology based on FEM modeling to assess plastic behavior of Cu has been developed (including crystal plasticity and data from OIM to reproduce the beam structure - grain size and orientation). The results agree with data reported on fcc metallic micro- and nano-pillars tested in compression and in tension.

ACKNOWLEDGMENTS

The authors gratefully acknowledge the financial support of Intel Corporation for the realization of this work. One of the authors (M. Trueba) wants to acknowledge the support given by the University of Navarra during her Doctoral Thesis.

REFERENCES

1. R.H. Dauskardt, M. Lane, Q. Ma and N. Krishna, *Engng. Fracture Mech.* **61**, 141-162 (1998).
2. J.M. Sánchez, S. El-Mansy, B. Sun, T. Scherban, N. Fang, D. Pantuso, W. Ford, M.R. Elizalde, J.M. Martínez-Esnaola, A. Martín-Meizoso, J. Gil-Sevillano, M. Fuentes and J. Maiz. *Acta Mater* **47**, 4405-4413 (1999).
3. I. Ocaña, J.M. Molina-Aldareguia, D. Gonzalez, M.R. Elizalde, J. Gil Sevillano, T. Scherban, D. Pantuso, B. Sun, G. Xu, B. Miner, J. He and J. Maiz, *Acta Mater* **54**, 3453–3462 (2006).
3. R. Huang, J.H. Prévost, Z.Y. Huang and Z. Suo, *Engng. Fracture Mech.* **70**, 2513-2526 (2003).
4. G.R. Anstis, P. Chantikul, B.R. Lawn, D.B. Marshall, *J Am Ceram Soc* **64**, 533-8 (1981).
5. D.S. Harding, W.C. Oliver and G.M. Pharr, *Mater. Res. Soc. Symp. Proc.*356-663 (1995).
6. D.J. Morris and R.F. Cook, *Int J Fract* **136**, 237-64 (2005).
7. A.K. Bhattacharya and W.D. Nix, *Int J Solids Struct* **27**, 1047-58 (1991).
8. K. Matoy, H. Schönkerr, T. Detzel and G. Dehm, *Thin Solid Films* **518**, 5796-5801 (2010).
9. D. González, "Numerical Tools for the Finite Element Modelling of Fracture and Crystal Plasticity", Ph.D. Thesis, University of Navarra, 2008.
10. J.R. Greer and J.T.M. De Hosson, *Progress in Materials Science* **56**, 654-724 (2011).

3D imaging of semiconductor components by discrete laminography

K.J. Batenburg[*,†], W.J. Palenstijn[†] and J. Sijbers[†]

[*]Centrum Wiskunde & Informatica, P.O. Box 94079, NL-1090 GB Amsterdam, The Netherlands
[†]iMinds-Vision Lab, University of Antwerp, Universiteitsplein 1, B-2610 Wilrijk, Belgium

Abstract. X-ray laminography is a powerful technique for quality control of semiconductor components. Despite the advantages of nondestructive 3D imaging over 2D techniques based on sectioning, the acquisition time is still a major obstacle for practical use of the technique. In this paper, we consider the application of *Discrete Tomography* to laminography data, which can potentially reduce the scanning time while still maintaining a high reconstruction quality. By incorporating prior knowledge in the reconstruction algorithm about the materials present in the scanned object, far more accurate reconstructions can be obtained from the same measured data compared to classical reconstruction methods. We present a series of simulation experiments that illustrate the potential of the approach.

Keywords: Laminography, discrete tomography
PACS: 81.70.Tx

INTRODUCTION

Precise quality control inspections of soldering and assembly of electronic devices have become priority items in the electronics manufacturing industry. Many inspection systems for such devices make use of X-rays to form images (radiographs) that exhibit features representing the internal structure of the devices and connections. Interpretation of a single radiograph, however, is not obvious as depth information is completely lost. In an attempt to compensate for this shortcoming, laminographic systems have been built in which the object is viewed from a plurality of angles. The additional views enable these systems to partially resolve the ambiguities present in the X-ray radiograph.

In Fig. 1, a laminographic setup is depicted. The X-ray beam is rotated with respect to a vertical axis. The magnification and field of view are controlled by projecting the beam at different angles with respect to the axis of rotation. The detector rotates at the same speed as the beam, but 180° out of phase. By this arrangement, a single focal plane is created through which the beam travels for the entire duration of one rotation.

The focal plane can be simply reconstructed by shifting and adding the constituent projection images to bring structures into registration. The focal plane at an arbitrary depth can be determined and reconstructed [1]. Although this traditional shift-and-add tomosynthesis is simple and computationally effective, it suffers from significant blurring artefact due to the superposition of objects from other planes. Indeed, radiograph contributions from out-of-plane material, noise, and rotation of the X-ray source around the object during each laminograph formation, all contribute to imperfections in the representation of the object within the focal plane. Therefore, the development of im-

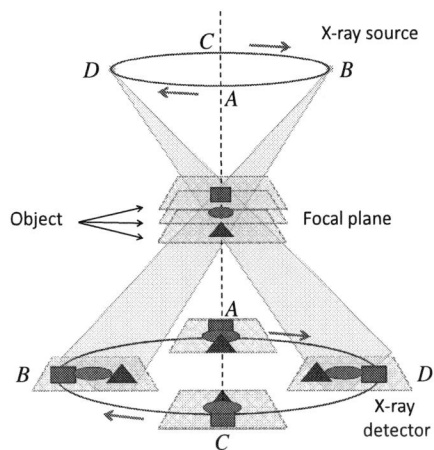

FIGURE 1. Laminography setup

proved reconstruction algorithms to correct for the out-of-plane blur is essential. Many approaches to reduce the blur artefact have been introduced [2, 3, 4]. The filtered back-projection approach, which consists of applying a filtering step to the projection data followed by a backprojection of the filtered data, is commonly used in practice. Iterative methods, such as the Simultaneous Iterative Reconstruction Technique (SIRT) result in fewer artefacts, at the expense of a longer reconstruction time [5]. Matrix Inversion Tomosynthesis (MITS) uses known imaging geometry and linear systems theory to deterministically separate in-plane detail from residual tomographic blur [6]. Despite these efforts, laminographic reconstruction still suffers from a limited depth resolution and various artefacts, mainly due to an inherent lack of a sufficient number of radiographs, which makes the reconstruction problem underdetermined.

The problem of reconstructing an image from a small number of radiographs has recently attracted considerable interest in the field of Compressed Sensing [7, 8]. In particular, it was proved that if the image is sparse, it can be reconstructed accurately from a small number of measurements with very high probability. In many images of objects that occur in practice, the image itself is not sparse, yet the boundary of the object is relatively small compared to the total number of pixels. In such cases, sparsity of the gradient image can be exploited by Total Variation Minimization [9, 10]. In this paper, we consider a different type of prior knowledge, where it is assumed that the object to be reconstructed consists of a small number (i.e., 2-5) of different materials, each corresponding to a characteristic, approximately constant gray level in the reconstruction. This type of reconstruction problem is known as *Discrete Tomography* (DT) [11]. Recently, the discrete algebraic reconstruction technique (DART), was introduced as an algorithm that can incorporate such prior knowledge [12]. The DART technique is an iterative reconstruction method that assumes the objects to be piecewise uniform with known densities. This reconstruction method has proved to be effective in reconstructing high quality images from only a limited number of radiographs [13, 14, 15].

In this paper, the use of DART for X-ray laminography of through-silicon Vertical

Interconnect Accesses (VIAs) is explored. We first provide an outline of the algorithmic ideas behind DART and discuss the particular obstacles that must be overcome when using DART for laminography applications. Subsequently, a series of simulation experiments is described that have been performed to assess the feasibility of using discrete tomography for the reconstruction of through-silicon VIAs from X-ray laminography data. The results of these experiments indicate that DART has strong potential for reducing the number of projection images required, and therefore the required scanning time.

METHODS

In this section, we outline the DART algorithm and discuss the specific considerations of applying DART to the reconstruction of copper through-silicon VIAs (TSVs) from X-ray laminography data. DART requires prior knowledge of the gray levels in the reconstructed image, which correspond to the different materials in the object. By effectively using this prior knowledge, a more accurate reconstruction can be obtained compared to algorithms that do not restrict the gray levels. For the experiments presented in this paper, we focus on the reconstruction of a simulated TSV for which the complete structure exhibits three gray levels, for void, silicon and copper.

As a starting point, we briefly review the iterative SIRT algorithm, which computes a grayscale reconstruction from a tomography (or laminography) dataset. A high level overview of the DART algorithm, which uses SIRT as a subroutine, is subsequently presented. The reader is referred to [12] for details about the implementation of DART.

The SIRT algorithm

The SIRT algorithm is a standard reconstruction technique for reconstructing grayscale images from a number of X-ray images, collected from different angles. It computes an approximate solution of the linear system $Wx = p$, where the vector x contains the gray level (also called pixel value) for each pixel, the vector p contains the measured projection data, and the matrix W represents the projection operation (i.e., computing the product Wx yields the projections corresponding to the image x). If no exact solution of this system exists, SIRT computes a solution for which the L^2 norm of the difference between the computed projection and the measured data, $Wx - p$ (also called projection error), is minimal, i.e., a least-squares solution (see [5] for details). Subsequently, each image pixel value is updated by adding a weighted average of the projection difference for all lines that intersect this pixel.

An important feature for using this algorithm within DART, is the fact that SIRT can also compute a grayscale reconstruction on a *subset* of the image pixels, where the intensity in the projections is only distributed among the selected subset. As we will see in the next section, DART uses this feature to focus the reconstruction on the object boundary.

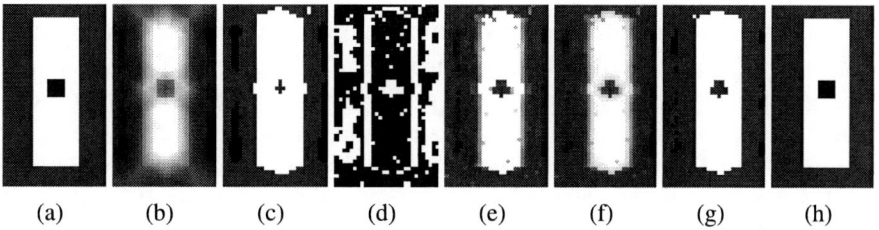

| (a) | (b) | (c) | (d) | (e) | (f) | (g) | (h) |

FIGURE 2. Illustrations of the different stages within an iteration of DART.

Overview of DART

In this section, we will introduce the DART algorithm and give an overview of the basic algorithmic steps. The DART algorithm is an iterative method, which starts from an initial image estimate and then repeatedly updates this image to improve the quality of the reconstruction.

As an example, suppose that we want to reconstruct a silicon sample containing a number of TSVs embedded in the silicon, some of which possibly containing voids. A cross-section of one of these TSVs is shown in Fig. 2(a).

Before applying DART, one needs an estimate of the number of gray levels in the reconstruction (i.e., the number of materials in the sample), as well as the actual gray levels. In this case we assume that three different densities are present: void, silicon and copper. An estimate of the actual graylevels can be hard to obtain, in particular when only a small number of projections are available. As a first step before applying DART, a conventional SIRT reconstruction is usually computed to obtain information about the materials in the sample and their gray levels. A SIRT reconstruction of the TSV is shown in Fig. 2(b).

Although the TSVs have a constant density, the SIRT reconstruction exhibits a spectrum of gray values. In addition, the shape of the reconstructed structure is clearly distorted, due to the limited angular range of the available projections. The range of gray levels in the reconstruction immediately presents a problem when the reconstructed image needs to be segmented. Segmentation is commonly performed by thresholding, but it is not obvious at all which threshold should be chosen in this case. The DART algorithm starts from a thresholded SIRT reconstruction, and then iteratively improves upon the current segmentation. Although a threshold also has to be chosen for DART, its choice is of minor importance to the final result. In this example, we choose the thresholds to be exactly in the middle between the gray level of the background and copper, and between copper and silicon, respectively. Fig. 2(c) shows the thresholded SIRT reconstruction. The thresholded reconstruction shows that pixels of the interior of the object that are not too close to the boundary are assigned the correct segmentation class (void, silicon, or copper). Pixels that are close to the boundary can be detected automatically from the thresholded SIRT reconstruction, by checking if any of the surrounding pixels belongs to a different segmentation class. We refer to these pixels as boundary pixels, and to the remaining pixels as non-boundary pixels. Fig. 2(d) shows the set of boundary pixels computed from Fig. 2(c). Note that even some non-boundary pixels have the wrong gray level in the thresholded SIRT reconstruction, when compared to the original TSV image.

We now turn back to the SIRT reconstruction in Fig. 2(b). The non-boundary pixels are assigned the gray level that corresponds to their respective material class (void, silicon, or copper). Next, the SIRT algorithm is used again, but only the *boundary* pixels are allowed to vary. The non-boundary pixels are kept fixed at their discrete levels. In this way, the number of variables in the linear equation system $Wx = p$ is significantly reduced, while the number of equations remains the same. Fig. 2(e) shows the result after 10 SIRT iterations for the boundary pixels. The SIRT step on the boundary can cause heavy fluctuations of the gray levels. A smoothing step is performed on the boundary to remove this roughness. The result is shown in Fig. 2(f). Applying a stronger smoothness filter leads to less noise and smoother boundaries in the reconstruction, while losing some fine single-pixel details. We now repeat the same steps as before, starting from the reconstruction in Fig. 2(f). The result of the threshold step is shown in Fig. 2(g). It is already very clear that the quality of the segmentation has improved considerably in just one iteration. By performing several iterations, the reconstruction quality is further improved, as shown in Fig. 2(h) for 10 full DART iterations.

DART for laminography

Compared to electron tomography, where the DART algorithm has already been applied successfully to the reconstruction of a range of nanomaterials [13, 14, 15], the laminography case is substantially more challenging from an implementation point-of-view. As the viewing directions for the series of projection images do not share a common plane, a fully-3D volume representation of the object is needed to compute the projections. Subdividing the volume into a stack of slices that are each processed independently is not possible in this case. As a consequence, the entire image volume must be kept in memory all at once, which imposes significant memory requirements.

Computing a series of projections for a given voxel object is computationally expensive. As each voxel contributes to each projection, the total number of mathematical operations that must be performed to compute a series of projections is proportional to the product of the number of voxels and the number of projection images. This computation can easily take several hours for a series of 200 projections of a $1024 \times 1024 \times 1024$ voxel volume, when run on a standard PC. In recent years, Graphics Processing Units (GPUs) have developed into highly parallel multi-processors that can process up to 1000 operations simultaneously. This high degree of parallelism is possible for the tomography computation, as the *same* operations must be performed to a large number of data elements.

In [16], we presented a GPU-based algorithm for the projection and backprojection operations that attains a speedup of two orders of magnitude compared to a conventional CPU implementation. By using the 3D version of this code for laminography reconstruction, the running time for a complete 3D DART reconstruction of a $500 \times 500 \times 50$ volume can be reduced to around 15 minutes on an NVIDIA GTX580 GPU. A key limitation of this GPU-based approach is the fact that for maximal efficiency, the entire voxel volume and the series of projections must be stored in the RAM memory on the

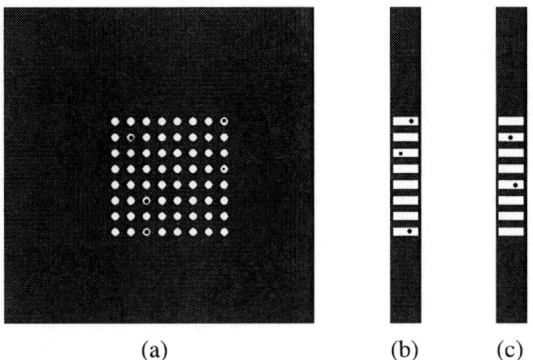

(a)   (b) (c)

FIGURE 3. Cross-sections of the simulation phantom.

graphics card itself. This memory is limited to 6GB–12GB for the high-end NVIDIA Tesla cards, limiting the size of volumes for which this high efficiency can be obtained.

EXPERIMENTS

To validate the use of DART for X-ray laminography of TSVs, a series of simulation experiments was set up. The experiments are based on a simulated 3D phantom object, which represents a slab of silicon containing a regular grid of copper TSVs, some of which contain voids. This results in a voxelized phantom containing three gray levels for the voids, silicon, and copper.

The phantom volume contains a stack of 50 slices, each consisting of 500×500 voxels, corresponding to a thin slab of silicon. The TSVs have a diameter of 11 voxels, whereas voids contained in these TSVs have a diameter of 6 voxels. Cross-sections of the phantom in three directions are shown in Fig. 3. The projection image recorded for each angle covers only a part of the silicon slab. As a result, only the central part of the slab is covered by all projections. Fig. 4(a-b) illustrate the coverage of different parts of the phantom by the set of projections, where the white region is covered by all angles and the black region is not covered at all. The experimental results presented in the next section focus on a particular Region-of-Interest (ROI) within the silicon slab that is covered by all angles. The extent of this region is depicted in Fig. 4(c-d).

Synthetic projection data of the phantom were generated by simulating a parallel beam laminography setup, as follows: First, the Radon transform of the phantom was computed, resulting in a series of projection images, where the width of a detector cell was taken to be the same as the voxel size of the phantom. Next, (noiseless) CT projection data were generated where a mono-energetic X-ray beam was assumed. The projections were then perturbed with Poisson distributed noise where the number of counts per detector element I_0 (flat field) was varied from 10^4 (high noise) up to 10^6 (low noise). Next, the linearized noisy projection data was obtained by dividing the CT projection data by the flatfield intensity and computing the negative logarithm. In this way, simulated projection images were obtained for varying signal-to-noise ratios. Finally, the simulated, noisy CT images were reconstructed using both the SIRT

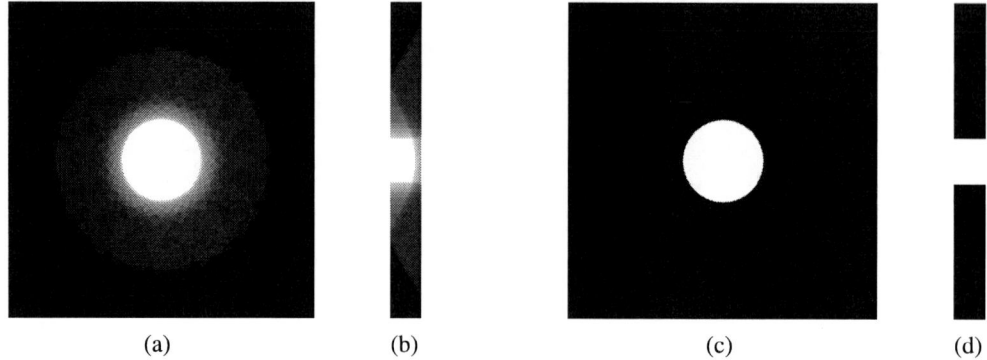

FIGURE 4. (a-b): Illustration of the coverage of the cross-section shown on the left by the series of projections. (c-d) Region-of-Interest that is covered by all projections.

algorithm and the DART algorithm.

The SIRT reconstructions were thresholded to obtain an image containing just three gray levels, using the well-known method of Otsu [17]. The resolution of the reconstruction depends strongly on the position within the reconstructed volume, varying both in depth and in the lateral position within the slab. In the presentation of the results we focus on two slices through the reconstructed volume, parallel to the silicon slab (xy-plane) and through the slab (xz-plane).

To compare the quality of the reconstructions, the *Number of Misclassified Pixels* (NMP) is used, which is defined as the number of misclassified pixels in the ROI of the segmented reconstruction.

RESULTS

Two series of experiments have been performed, comparing the reconstruction quality of SIRT and DART, where the SIRT algorithm is considered as a representative method from the current class of "relatively advanced" methods applied in X-ray laminography. The first set of experiments is aimed at investigating the dependency of the reconstruction quality on the number of projection images, while the second set is aimed at investigating the influence of noise.

Varying the number of projections

In a first series of experimental results, we provide a qualitative impression of the reconstruction quality obtained by both SIRT and DART, when varying the number of projection images. Fig. 5 shows reconstructed cross-sections through the silicon in two directions, where the number k of projections is varied from $k = 5$ up to $k = 60$.

From the results shown in Fig. 5, it is clear that within the region of silicon that is covered by all projections, DART allows to reduce the number of projections down to 10 while still obtaining reconstructions of good visual quality.

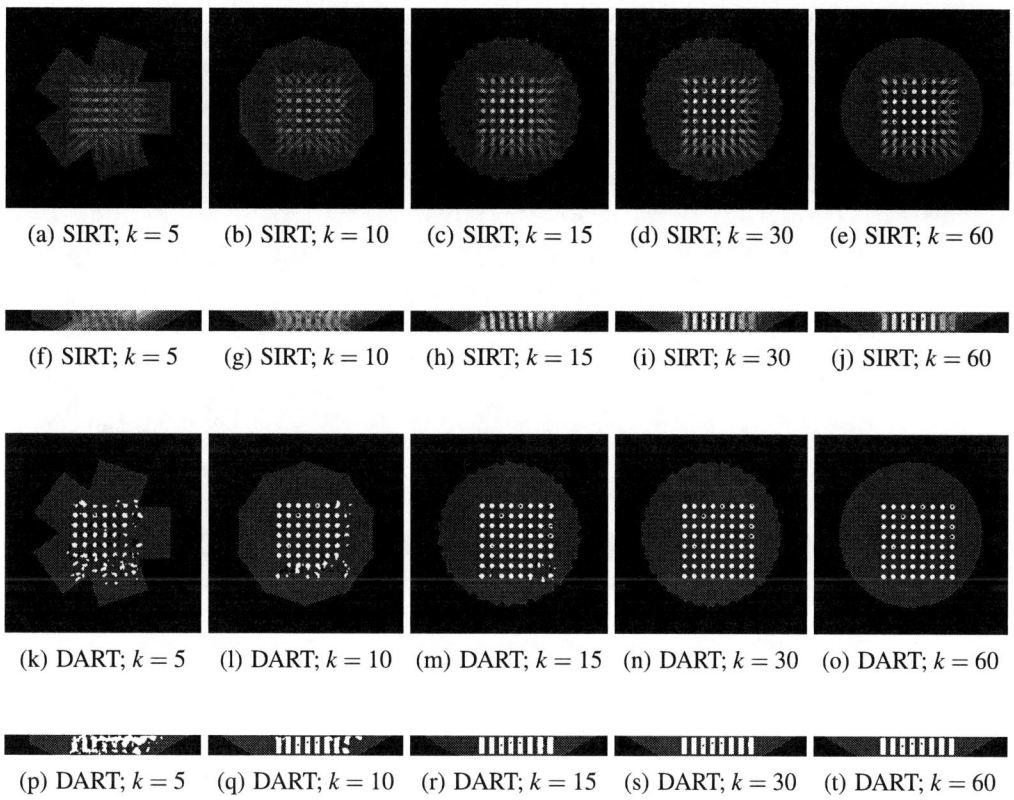

(a) SIRT; $k = 5$ (b) SIRT; $k = 10$ (c) SIRT; $k = 15$ (d) SIRT; $k = 30$ (e) SIRT; $k = 60$

(f) SIRT; $k = 5$ (g) SIRT; $k = 10$ (h) SIRT; $k = 15$ (i) SIRT; $k = 30$ (j) SIRT; $k = 60$

(k) DART; $k = 5$ (l) DART; $k = 10$ (m) DART; $k = 15$ (n) DART; $k = 30$ (o) DART; $k = 60$

(p) DART; $k = 5$ (q) DART; $k = 10$ (r) DART; $k = 15$ (s) DART; $k = 30$ (t) DART; $k = 60$

FIGURE 5. Cross-sections of the reconstructed volume computed by SIRT and DART from a varying number of projections; First row: xy-slice, SIRT; Second row: xz-slice, SIRT; Third row: xy-slice, DART; Fourth row: xz-slice, DART.

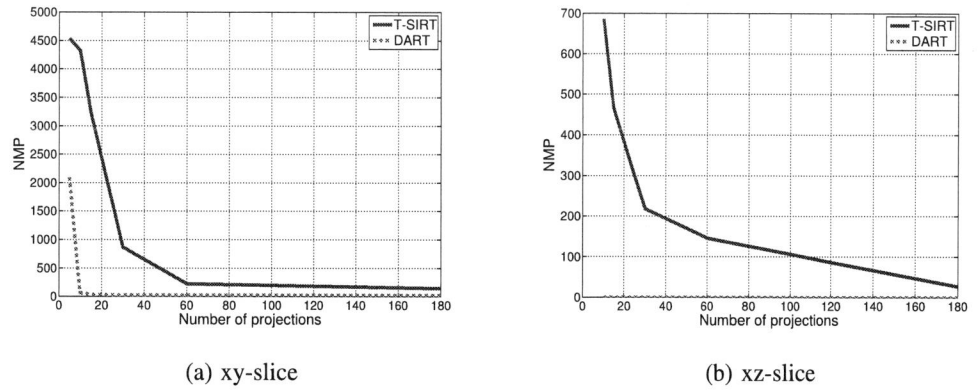

(a) xy-slice

(b) xz-slice

FIGURE 6. Number of Misclassified Pixels (NMP) in the reconstructed ROI for a varying number of projections, for both SIRT and DART. Left: error in xy-slice; Right: error in xz-slice.

In Fig. 6, a *quantitative* comparison is made between the SIRT and DART reconstructions, plotting the NMP in the ROI as a function of the number of projections. Fig. 6a shows the NMP in a slice parallel to the silicon slab (xy-slice), while Fig. 6b shows the results for a slice in the direction of the VIAs (xz-slice). It can be observed that even when using only 10 projections, the error for DART already drops to a level which is not attained by SIRT, even when using as many as 180 projections.

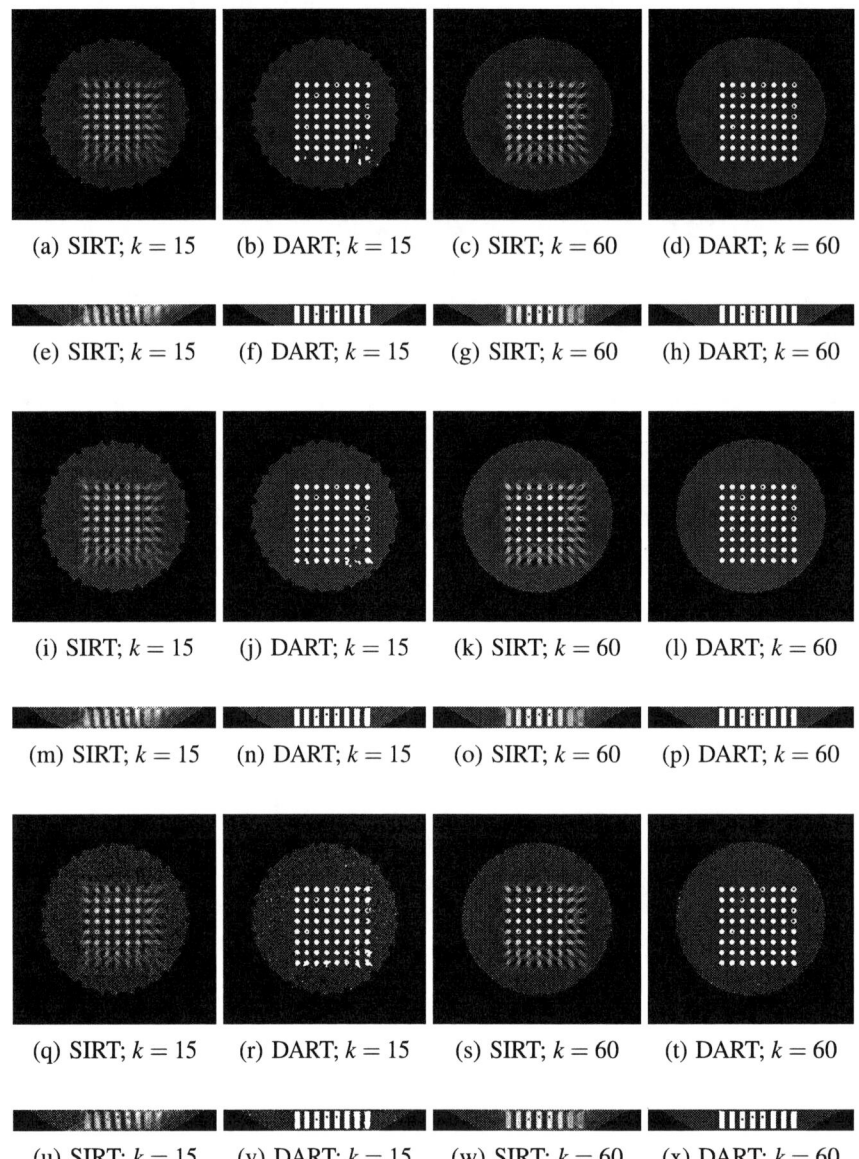

    (a) SIRT; $k = 15$ (b) DART; $k = 15$ (c) SIRT; $k = 60$ (d) DART; $k = 60$

    (e) SIRT; $k = 15$ (f) DART; $k = 15$ (g) SIRT; $k = 60$ (h) DART; $k = 60$

    (i) SIRT; $k = 15$ (j) DART; $k = 15$ (k) SIRT; $k = 60$ (l) DART; $k = 60$

    (m) SIRT; $k = 15$ (n) DART; $k = 15$ (o) SIRT; $k = 60$ (p) DART; $k = 60$

    (q) SIRT; $k = 15$ (r) DART; $k = 15$ (s) SIRT; $k = 60$ (t) DART; $k = 60$

    (u) SIRT; $k = 15$ (v) DART; $k = 15$ (w) SIRT; $k = 60$ (x) DART; $k = 60$

FIGURE 7. Cross-sections (xy-slice) of the reconstructed volume computed by SIRT and DART from varying noise levels; Row 1-2: no noise; Row 3-4: moderate noise ($I_0 = 100000$); Row 5-6: high noise ($I_0 = 10000$).

Varying the noise level

In a second series of experiments, the robustness of DART is compared to SIRT for varying noise levels in the projection data, for two sets of projection angles ($k = 15$ and $k = 60$). A qualitative impression of the reconstruction results is shown in Fig. 7.

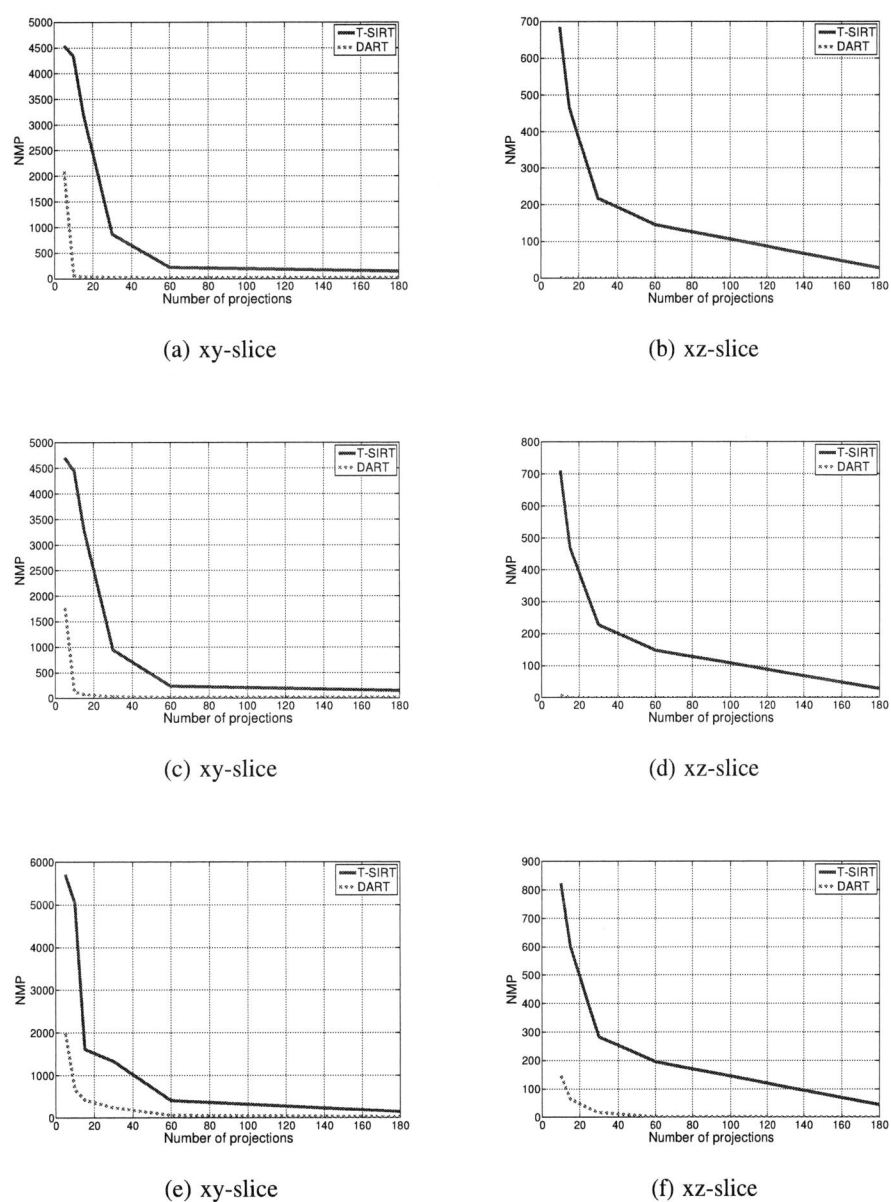

FIGURE 8. Number of Misclassified Pixels (NMP) in the reconstructed ROI for a varying number of projections, for both SIRT and DART. Left: error in xy-slice; Right: error in xz-slice. Row 1: no noise; Row 2: moderate noise; Row 3: high noise.

The results in Fig. 7 illustrate that even if the projection data contains a fair amount of noise, the ROI can still be reconstructed accurately using DART, even from a low number of $k = 15$ projections.

Fig. 8 shows a quantitative comparison of SIRT and DART for a varying number of angles and for three different noise levels: no noise, moderate noise ($I_0 = 100000$) and high noise ($I_0 = 10000$). In all cases, DART clearly outperforms SIRT in terms of both the NMP and the visual quality of the reconstruction.

DISCUSSION AND CONCLUSIONS

The results presented in this paper provide a first indication that discrete tomography could be successfully applied to the reconstruction of electrical components from X-ray laminography measurements. To assess the true practical value of the technique in this domain, the DART algorithm must now be tested on experimental datasets. These real-world experiments will include the effects of mechanical instabilities, beam-hardening and other physical sources of error that cause a mismatch with the idealized linear projection model employed in DART.

However, given the fact that DART has already proven itself on experimental data in several domains, including the highly challenging domain of electron tomography, we are positive about the prospects of such experiments, which may eventually lead to a strong reduction in the number of projections required for a full quality assessment, and therefore in a reduction of the scanning time.

ACKNOWLEDGMENTS

KJB acknowledges support from the Netherlands Fund for Scientific Research (NWO), project number 639.072.005.

REFERENCES

1. Z. Kolitsi, G. Panayiotakis, V. Anastassopoulos, A. Scodras, and N. Pallikarakis, *Medical Physics* **19**, 1045–1050 (1992).
2. D. N. Ghosh Roy, R. A. Kruger, B. Yih, and P. Del Rio, *Medical Physics* **12**, 65–70 (1985).
3. D. P. Chakraborty, M. V. Yester, G. T. Barnes, and A. V. Lakshminarayana, *Radiology* **150**, 225–229 (1984).
4. Z. Kolitsi, G. Panayiotakis, and N. Pallikarakis, *Medical Physics* **20**, 47–50 (1993).
5. J. Gregor, and T. Benson, *IEEE Trans. Image Proc.* **27**, 918–924 (2008).
6. J. T. Dobbins III, and D. J. Godfrey, *Physics in Medicine and Biology* **48**, R65–R106 (2003).
7. D. L. Donoho, *IEEE Transactions on Information Transfer* **52**, 1289–1306 (2006).
8. E. Y. Sidky, C.-M. Kao, and X. Pan, *J. X-ray Sci. Tech.* **14**, 119–139 (2006).
9. A. Chambolle, *Journal of Mathematical Imaging and Vision* **20**, 89–97 (2004).
10. E. Y. Sidky, and X. Pan, *Phys. Med. Biol.* **53**, 4777–4807 (2008).
11. G. T. Herman, and A. Kuba, editors, *Discrete Tomography: Foundations, Algorithms and Applications*, Birkhäuser, Boston, 1999.
12. K. J. Batenburg, and J. Sijbers, *IEEE Transactions on Image Processing* **20**, 2542–2553 (2011).
13. S. Bals, K. J. Batenburg, J. Verbeeck, J. Sijbers, and G. Van Tendeloo, *Nano Letters* **7**, 3669–3674 (2007).

14. K. J. Batenburg, S. Bals, J. Sijbers, C. Kübel, P. A. Midgley, J. C. Hernandez, U. Kaiser, E. R. Encina, E. A. Coronado, and G. Van Tendeloo, *Ultramicroscopy* **109**, 730–740 (2009).
15. T. Roelandts, K. J. Batenburg, E. Biermans, C. Kübel, S. Bals, and J. Sijbers, *Ultramicroscopy* **26**, 96–105 (2012).
16. W. J. Palenstijn, K. J. Batenburg, and J. Sijbers, *Journal of structural biology* **176**, 250–253 (2011).
17. N. Otsu, *IEEE Trans. Systems, Man, and Cybernetics* **9**, 62–66 (1979).

Effects of Fluoride Residue on Thermal Stability in Cu/Porous Low-k Interconnects

Y. Kobayashi, S. Ozaki, Y. Nakata and T. Nakamura

FUJITSU LABORATORIES Ltd., 10-1 Morinosato-Wakamiya, Atsugi, Kanagawa 243-0197, Japan

Abstract. We have investigated the effects of fluoride residue on the thermal stability of a Cu/barrier metal (BM)/porous low-k film (k < 2.3) structure. We confirmed that the Cu agglomerated more on a BM/inter layer dielectric (ILD) with a fluoride residue. To consider the effect of fluoride residue on Cu agglomeration, the structural state at the Cu/BM interface was evaluated with a cross-section transmission electron microscope (TEM) and atomic force microscope (AFM). In addition, the chemical bonding state at the Cu/BM interface was evaluated with the interface peeling-off method and X-ray photoelectron spectroscopy (XPS). Moreover, we confirmed the ionization of fluoride residue and oxidation of Cu with fluoride and moisture to clarify the effect of fluoride residue on Cu. Our experimental results indicated that the thermal stability in Cu/porous low-k interconnects was degraded by enhancement of Cu oxidation with fluoride ions diffusion as an oxidizing catalyst.

Keywords: Thermal stability, Cu agglomeration, Low-k, Barrier metal oxidation, Fluoride residue, Ionization, diffusion
PACS: 68

INTRODUCTION

It is well recognized that a combination of Cu wiring and a low-k inter layer dielectric (ILD) is required to reduce the resistance-capacitance delay of LSI interconnects.

However, introducing low-k ILDs lowers the reliability of interconnects. It has been reported that moisture absorption in a low-k ILD, which involves barrier metal (BM) oxidation, could be a possible origin of this reliability degradation [1-3]. Similarly, fluoride residues such as CHxFy-polymers and F-diffused layers on low-k ILD that are generated during ILD etching [4, 5] can also increase BM oxidation, as mentioned in our previous report [6].

In this study, we have investigated the effects of fluoride residue on the thermal stability of a Cu/BM/ILD structure.

EXPERIMENTAL

Silica-based spin-on porous dielectrics (SOD) with k < 2.3 were used as ILDs for this study. One-hundred and fifty-nm-thick blanket SOD films were prepared on Si substrates. Some were etched by CF_4 plasma (as a sample with ILD etching), while other were not (as a sample w/o ILD etching). The amount of moisture was kept constant before and after etching to clarify the effects of the fluoride residue. The 10-nm-thick Ta barrier metal film and the subsequent 20-nm-thick Cu film were deposited on the ILD substrates with a radio frequency (RF) sputtering system. Figure 1 shows a schematic cross-section of a blanket structure used in this study. The samples' interfaces were analyzed by XPS and annealed under a vacuum for 5 minutes at 250°C for agglomeration testing. Sample interfaces for the XPS measurements were prepared with a peeling-off method [6] and without atmospheric exposure (Fig. 2).

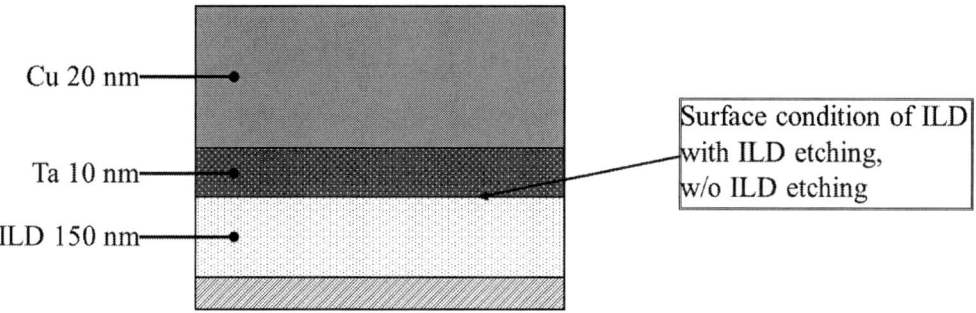

FIGURE 1. Schematic cross section of Cu/BM/ILD blanket structure. To investigate fluoride residue, two types of ILD surface conditions (with ILD etching and w/o ILD etching) were applied.

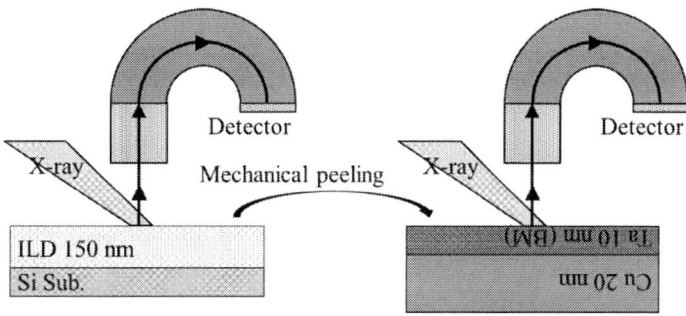

FIGURE 2. Peeling-off method for analysis at interfaces by XPS.

Additionally, for testing chemical stability of fluoride residue generated by ILD etching, an etched ILD film was dipped for an hour in boiling water at 100°C. After that, the boiled ILD film was analyzed by XPS, and post-boiling water was measured by ion chromatography and pH mater.

Next, for acceleration condition testing under simulated conditions to clarify the effect of fluoride on Cu, 20-nm-thick Cu films were prepared. Some were exposed to CF_4 plasma (as a sample of Cu with fluoride), while others were not (as a sample of Cu w/o fluoride). These films were prepared on a Si substrate and placed under accelerated conditions, which were 30°C and 60% relative humidity for 24 hours. After that, samples were analyzed by XPS.

RESULTS AND DISCUSSION

Structural investigation of Cu/BM interface

To evaluate the thermal stability of Cu wiring on BM/ILD with etching, simple annealed structure samples, which were the w/o-ILD-etching sample and with-ILD-etching sample, were observed under a scanning electron microscope (SEM). Figures 3a and 3b show SEM images of the Cu surfaces after annealing. Judging by the area in which the Ta surface appeared in Fig. 3b, the Cu agglomeration of the with-ILD-etching sample was determined to be larger than that of the w/o-ILD-etching sample. This means that the fluoride residue on the ILD surface accelerated the Cu agglomeration.

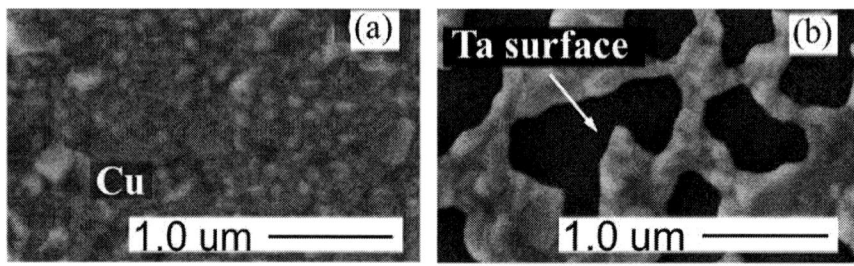

FIGURE 3. SEM images of Cu surface after annealing: (a) w/o ILD etching; (b) with ILD etching.

Now the question is why the fluoride residue accelerates Cu agglomeration at a distant location, which is the ILD surface. To answer this, the structural state after annealing was examined. Figures 4a and 4b show TEM images of the with-ILD-etching sample under each annealing condition. Although an obvious agglomeration of Cu is shown in Fig. 4b, the Ta coverage does not degrade.

Figure 4c shows the surface roughness of Ta after annealing by AFM. The surface roughnesses of the w/o-ILD-etching sample and with-ILD-etching sample were similar. Therefore, these investigations showed that fluoride residue does not lead to structural degradation.

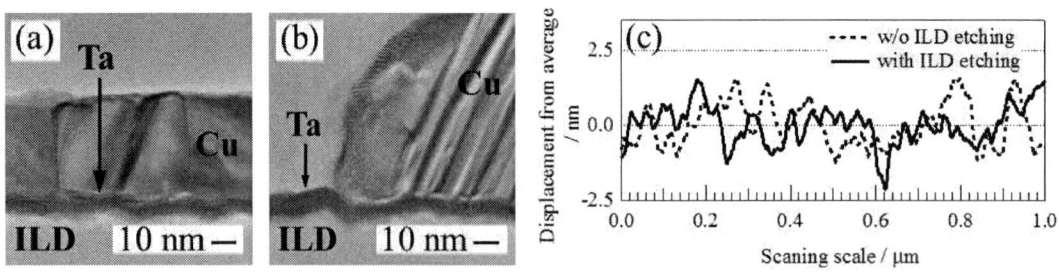

FIGURE 4. Cross-sectional TEM images of with-ILD-etching sample after annealing: (a) 200°C, 5 min.; (b) 250°C, 5 min. (c) Surface roughness of Ta after annealing by AFM.

Analysis chemical bonding state at Cu/BM and BM/ILD interface

To evaluate the change in chemical bonding state of the Cu/Ta interface due to fluoride residue, the Cu and Ta sides of the Cu/Ta/ILD structure were separated by mechanical peeling without being exposed to the air and subsequently analyzed by XPS. Table 1 lists the oxide ratio of Ta and Cu calculated by waveform separation of XPS and XAES [7] spectra at the Cu/Ta interface, respectively. As shown in Table 1, Cu oxidation of the with-ILD-etching sample significantly increased with near-complete Ta oxidation. This means the fluoride residue significantly increases not only Ta oxidation but also Cu oxidation at the Cu/Ta interface.

TABLE 1. Oxide ratio of Ta and Cu calculated by waveform separation of Ta 4f and Cu LMM spectra

Sample name	Rate (%)					
	Ti side			Cu side		
	Ta	Ta_2O_5	TaO_x	Cu	Cu_2O	CuO
W/o-ILD-etching sample	19.0	59.6	21.4	89.8	2.7	7.5
With-ILD-etching sample	1.3	77.7	20.9	21.8	59.2	18.9

Additionally, on Ta/ILD interface, Ta fluoride (684.8 eV) [9, 10] was detected in the Ta barrier metal of with-ILD-etching sample as shown in Fig. 5. Moreover, a fluoride peak is not detected at the ILD surface, where it would be normally. Thus, this suggests that fluoride residue diffused into barrier metal as fluoride ions.

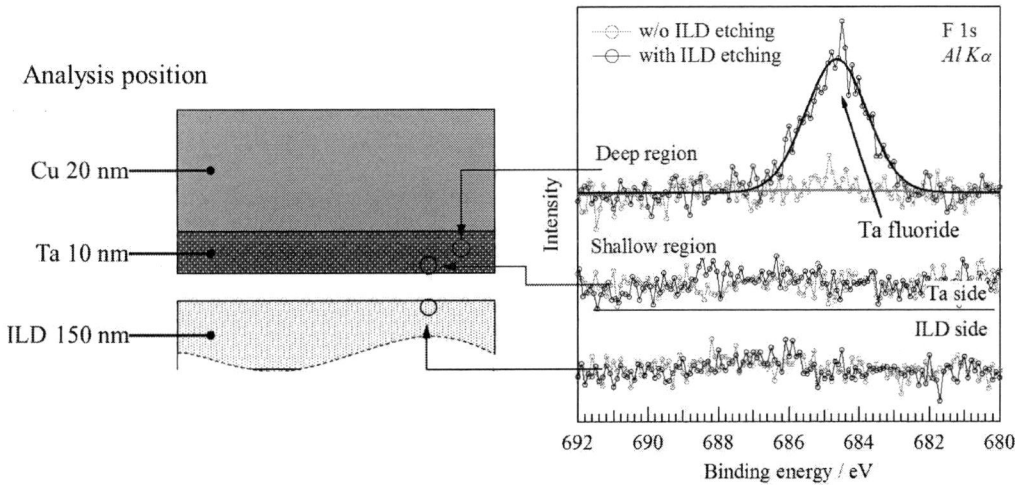

FIGURE 5. F 1s XPS spectra of Ta/ILD interface

Effects of fluoride residue on Cu agglomeration

To investigate the ionization of fluoride residue, etched ILD was dipped in boiling water and analyzed before and after boiling by XPS. Figure 6 shows the F 1s XPS spectra of etched ILD sample before and after boiling. The decrease in SiFx (686.0 eV) [11] was attributed to the hydrolysis reaction. On the other hand, no decrease in

CFx (686.8–690.3 eV) [9, 10] was observed. In addition, the boiling water after the dipping etched ILD sample was investigated by ion chromatography and pH measurement. As shown in Table 2, fluoride ions were detected in the water, and a pH decreased as fluoride ions concentration increased.

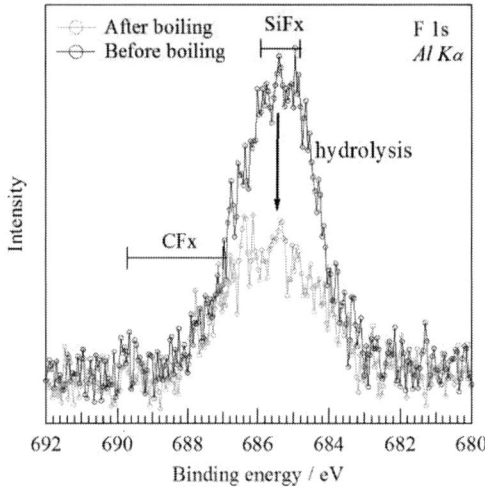

FIGURE 6. F 1s XPS spectra of etched ILD surface before and after boiling.

TABLE 2. Fluoride ions concentration and pH of boiling water after the dipping of etched ILD sample.

Condition	Fluoride ion concentration (ppb)	pH
Before boiling etched ILD	< 5 (below detection limit)	6.6
After boiling etched ILD	240	5.0

This tendency is reasonable from the viewpoint of the solution chemistry. Therefore, it is considered that SiFx reacted with water to form HF as follows [6]:

$$\equiv Si - F + H_2O \rightarrow \equiv Si - OH + HF, 2 \equiv Si - F + H_2O \rightarrow -Si - O - Si \equiv +2HF. \quad (1)$$

Next, to evaluate the effect of fluoride on Cu, the Cu-w/o-fluoride sample and Cu-with-fluoride sample were humidified under accelerated conditions and analyzed by XPS. Table 3 shows the atomic ratio of the oxidized Cu and fluoride calculated by XPS waveform separation. Although the atomic ratio of oxidized Cu was similar in both samples before humidifying, there was greater Cu oxidation of the Cu-with-fluoride sample. In addition, the amount of fluoride in the Cu-with-fluoride sample decreased after humidifying. This means that the fluoride on the Cu surface was involved in the Cu oxidation with moisture.

TABLE 3. Atomic ratio of oxidized Cu and fluoride calculated by XPS waveform separation.

Sample name	Atomic ratio (at.%)			
	Before humidifying		After 30°C, 60%RH 24h	
	Oxidized Cu	Fluoride	Oxidized Cu	Fluoride
Cu-w/o-fluoride sample	22.0	0.0	40.0	0.0
Cu-with-fluoride sample	23.6	29.2	59.6	1.8

Thus, it is suggested that fluoride residue generated by ILD etching was ionized by hydrolysis at BM/ILD interface, fluoride ions diffused into barrier metal, Cu at Cu/BM interface was oxidized by fluoride ions as oxidizing catalyst, and Cu agglomeration was enhanced.

CONCLUSIONS

We clarified that the Cu agglomerated more on a BM/porous low-k ILD with a fluoride residue. Although the structural state, which had the coverage and surface roughness of a barrier metal, did not change after Cu agglomeration, a significant increase in Cu oxidation with near-complete barrier metal oxidation and the existence of fluoride ions were confirmed in the barrier metal. Additionally, the fluoride residue generated by ILD etching was hydrolysable, and the Cu-with-fluoride residue increased the oxidation by humidifying the sample. These results indicate that the fluoride residue generated by ILD etching diffused into barrier metal as fluoride ions with moisture and increased the Cu oxidation significantly. Similarly, any fluoride residue in actual Cu/low-k interconnects will increase the Cu oxidation and degrade thermal stability.

REFERENCES

1. A. Sakata, S. Yamashita, S. Omoto, M. Hatano, J. Wada, K. Higashi, H. Yamaguchi, T. Yosho, K. Imamizu, M. Yamada, M. Hasunuma, S. Takahashi, A. Yamada, T. Hasegawa, H. Kaneko, "Reliability Improvement by Adopting Ti-barrier Metal for Porous Low-k ILD Structure," in Proceedings of the 2006 IEEE International Interconnect Technology Conference, San Francisco, 2006, pp. 101-104.
2. N. Matsunaga, N. Nakamura, K. Higashi, H. Yamaguchi, T. Watanabe, K. Akiyama, S. Nakao, K. Fujita, H. Miyajima, S. Omoto, A. Sakata, T. Katata, Y. Kagawa, H. Kawashima, Y. Enomoto, T. Hasegawa, H.Shibata, "BEOL Process Integration Technology for 45nm Node Porous Low-k/Copper Interconnects," in Proceedings of the 2005 IEEE International Interconnect Technology Conference, San Francisco, 2005, pp.6-8.
3. M. Hamada, K. Ohmori, K. Mori, E. Kobori, N. Suzumura, R. Etou, K. Maekawa, M. Fujisawa, H. Miyatake, A. Ikeda, "Highly reliable 45-nm-half-pitch Cu interconnects incorporating a Ti/TaN multilayer barrier," in Proceedings of the 2010 IEEE International Interconnect Technology Conference, San Francisco, 2010, pp.1-3.
4. Y. Furukawa, R. Wolters, H. Roosen, J. H. M. Snijders, R. Hoofman, Microelectron. Eng. 76, 25-31 (2004).
5. Y. Iba, T. Kirimura, M. Sakai, Y. Kobayashi, Y. Nakata, M. Nakaishi, Jpn. J. Appl. Phys. 47, 6923-6930 (2008).
6. S. Ozaki, Y. Nakata, Y. Kobayashi,T. Nakamura, Y. Iba, S. Fukuyama, H. Watatani, Y. Ohkura, Microelectron. Eng. 87, 370-372 (2010).
7. S. Poulston, P. M. Parlett, P. Stone, M. Bowker, Surf. Interface Anal. 24, 811-820 (1996).
8. J. P. Chang, H. W. Krautter, W. Zhu, R. L. Opila, C. S. Pai, J. Vac. Sci. Technol. A 17, 2969-2974 (1999).
9. A. Tressaud, E. Durand, C. Labrugère, J. Fluorine Chem. 125, 1639-1648 (2004).
10. G. Nansé, E. Papirer, P. Fioux, F. Moguet, A. Tressaud, Carbon 35, 175-194 (1997).
11. T. Takahagi, A. Ishitani, H. Kuroda, Y Nagasawa, J. Appl. Phys. 69, 803-807 (1991).

AIP Publishing LLC
Suite 1N01
2 Huntington Quadrangle
Melville, New York 11747

ISSN: 0094-243X
ISBN 978-1-63266-741-0

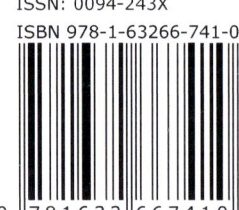